미래강소도시

Seoul National University IPUD
2024-1 Urban Design Studio 1
with TAEJAE FOUNDATION

미래강소도시
The robust compact city of the future

SEOUL NATIONAL UNIVERSITY
IPUD 협동과정 도시설계학전공
SNU - TAEJAE FOUNDATION
2024-1 Urban Design Studio 1

John Hong

STUDENTS

TEAM 1
Inseo Song
Sojung Yoon (TA)
Yeji Byun

TEAM 2
Halim Cha
Woojin Cho
Youngbin Im

TEAM 3
Hyeonguk Jeon
Inseo Cho
Soyun Park

TEAM 4
Charlaine Mae Nilsen
Sangwoo Park
Youngeun Moon

EDITOR AND DESIGNER
Sojung Yoon (TA)
Inseo Song
Yeji Byun
Sangwoo Park

This studio was made possible through the generous support of TAEJAE Foundation

TABLE OF CONTENTS

미래강소도시 : The Robust Compact City of the Future

In today's rapidly urbanizing world, the problems of large cities, such as congestion, inequality, and massive energy consumption, remind us of the need for sustainable cities of the future. Against this backdrop, 'compact cities', the opposite of megacities, are gaining attention as an environmentally friendly and livable urban model. In particular, the concept of the '15-minute city' aims for a city where all essential facilities are located within a 15-minute walk, emphasizing walkable streets, accessibility to everyday services, large parks and open spaces, and the integration of a circular economy.

The exhibition showcases the work of 12 students, divided into four teams, who were given a two-kilometer radius of land on which to design a future Robust Compact city. The land was given as a blank slate, and the students were given creative freedom to realize their own vision of the future city.

도시화가 급속도로 진행되는 현대 사회에서, 대도시의 문제점인 혼잡, 불평등, 대규모 에너지 소비 등의 문제는 우리에게 지속 가능한 미래 도시의 필요성을 일깨워줍니다. 이러한 배경 속에서 대도시와는 반대되는 개념인 '컴팩트 시티'는 환경 친화적이고 생활 편의성이 높은 도시 모델로 주목받고 있습니다. 특히, '15분 도시'라는 개념은 모든 생활 필수 시설이 도보 15분 내에 위치해 있는 도시를 지향하며, 걷기 좋은 거리, 일상적인 서비스 접근성, 넓은 공원 및 오픈 스페이스, 순환 경제의 통합 등을 강조합니다.

이번 전시는 12명의 학생들이 4개의 팀으로 나누어, 2km 반경의 대지를 바탕으로 미래의 강소도시를 설계한 결과물을 선보입니다. 주어진 대지는 아무것도 없는 상태였으며, 학생들은 창의적이고 자유롭게 자신만의 미래 도시 비전을 구현해냈습니다.

GRAVICiTY

Gravicity

1조 : 변예지 윤소정 송인서

There are two characteristics that cause all of the problems that humanity is currently facing: 1. Sustainability and 2. Inequality. However, it is very difficult to address the inequality between countries, inequality due to territorial size and resources, carbon emissions, and unsustainable development within a state system that is based on the boundary of a territory.

Current Problems?

State

UNSUSTAINABILITY

Carbon emissions
Reckless development

INEQUALITY

Size of the territory
Reckless development

New Solutions!

Cities

SUSTAINABILITY

CO2 Balance
Energy efficiency

EQUALITY

No boundaries
Accessibility

If we are given a new earth like white drawing paper, we will replace the current state's role with "the connection of sustainable cities."

**NO MORE STATE!
MORE CITIES!**

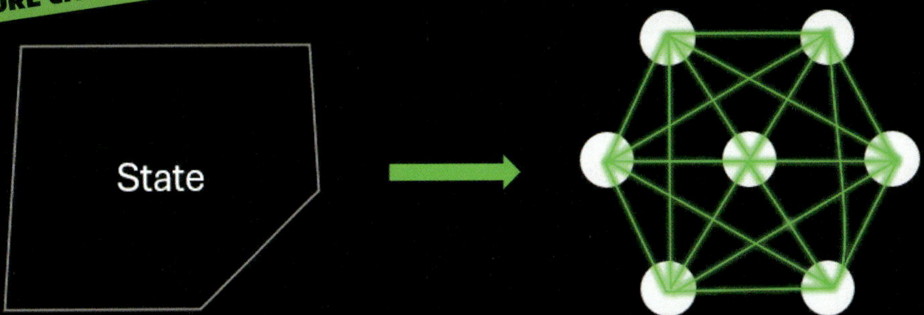

State based on boundary of its territory

Union of sustainable cities

An essential factor in creating a sustainable city is the regulation of eco-friendly carbon. We want to achieve carbon zero by balancing the carbon absorption and emissions of the new city. In order to balance these carbon, we adjust the density of the city and present new means of transportation.

How can we protect the environment?

Balance the CO2 emissions in cities!

URBAN DENSITY

NEW MOBILITY

Gravity Grid

Gravity Grid is a grid of cities designed to control the density of cities.

The Gravity Grid is a grid conceived of a heavy object placed on a cloth as follows. The grid has a dense center and its outer periphery is widened. The inner block pressed by the heavy ball decreases in size and the outer block increases in size, making the center of the city the densest block. It is a grid suitable for increasing the density inside and lowering the density outside the city.

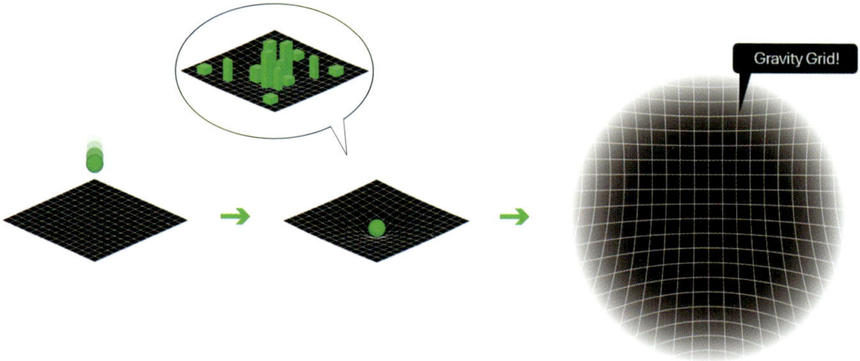

Gravity Grid_How does it function?

The reason for the density control for the carbon balance is to achieve carbon zero by controlling the amount of carbon emissions in the city through building density control and matching it with the amount of carbon absorption. To this end, Gravity Grid performs five functions.

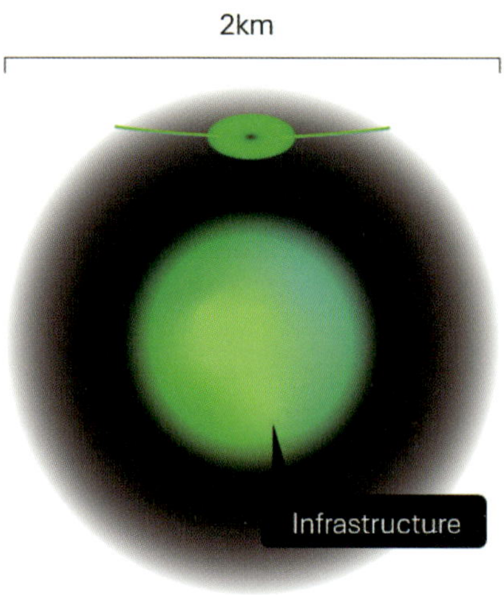

1. Limitation of Infrastructure

The infrastructure zone has a radius of 500m so that the area of the carbon emission source is not increased indefinitely. In addition, the formation of arbitrary large-scale infrastructure zones is restricted by limiting the distance between infrastructure zones to 10 km or more. This prevents the urban sprawl that causes indiscriminate carbon emissions.

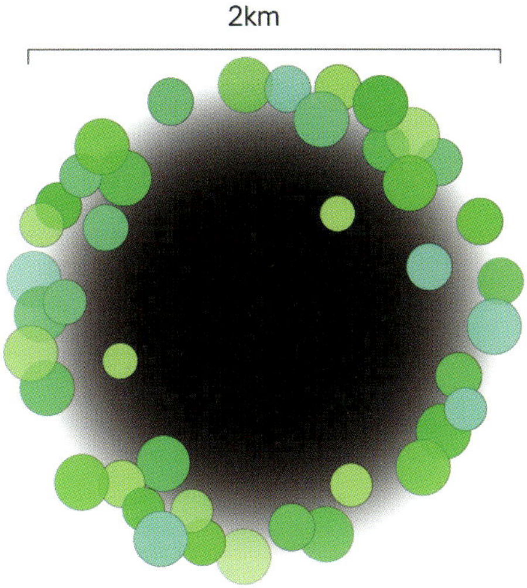

2. City Forest

It secures urban forests that are large enough to cover all of the carbon generated in the city. This contributes to zero carbon.

3. Grid changes according to various climates

The Gravity grid can be customized in terms of curvature, width, and block size through adjustments according to various climates, which contributes to reducing energy consumption. For example, in extremely cold climates, the curvature is assigned to control the amount of wind and the size of the block is increased to minimize heat loss.

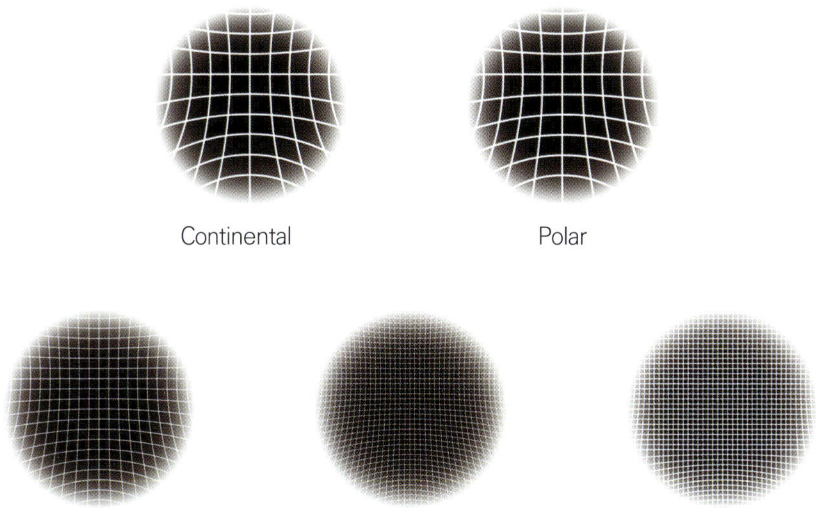

Continental

Polar

Temperate

Dry

Tropical

4. Gradation of Building Density

As you go to the outskirts of the city, you reduce carbon emissions by lowering the density of buildings and increasing green space. The increase in green space in the outskirts of the city could act as a buffer to separate outside nature from human activities in the center of the city.

5. Wheel System + Green Road

The range of the city is set to 2KM, which is walkable and therefore no vehicle.
All of the traffic has been replaced by new types of transportation, such as wheels and bicycles, and can be moved along the gravity grid. The Gravity grid is made of rust, so it absorbs carbon and is walk-friendly.

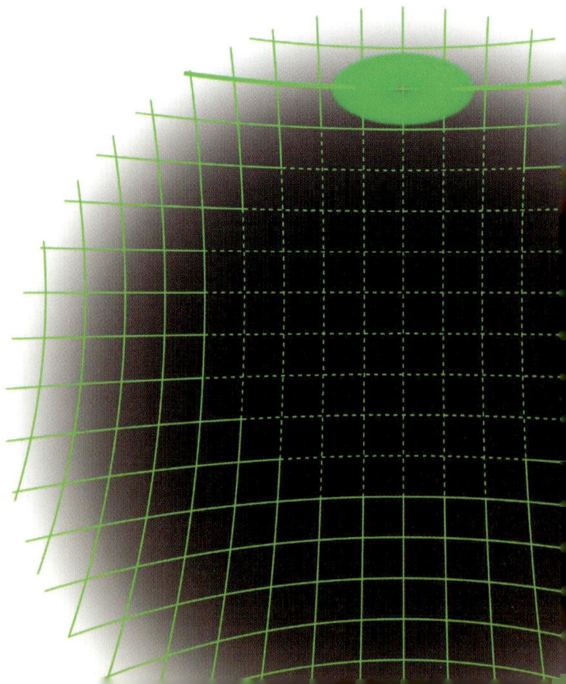

New Mobilty

One person emits about 5 tons of carbon emissions per year from traffic. Let me introduce the wheel, a new means of transportation designed to reduce this.
Wheel is self-driving and uses solar energy and kinetic energy after storing it.A wheel consists of an outer wheel and an inner wheel. A wheel swing bolt is installed here to store kinetic energy. The outer wheel is made of solar panels.

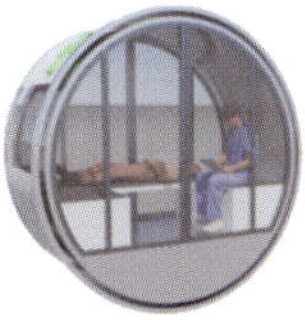

Medical Wheel

Ladder Wheel

Fire Wheel

Wheels are normally used for family and personal transportation. However, depending on the situation, it can be transformed into a medical, ladder, or fire Wheel.

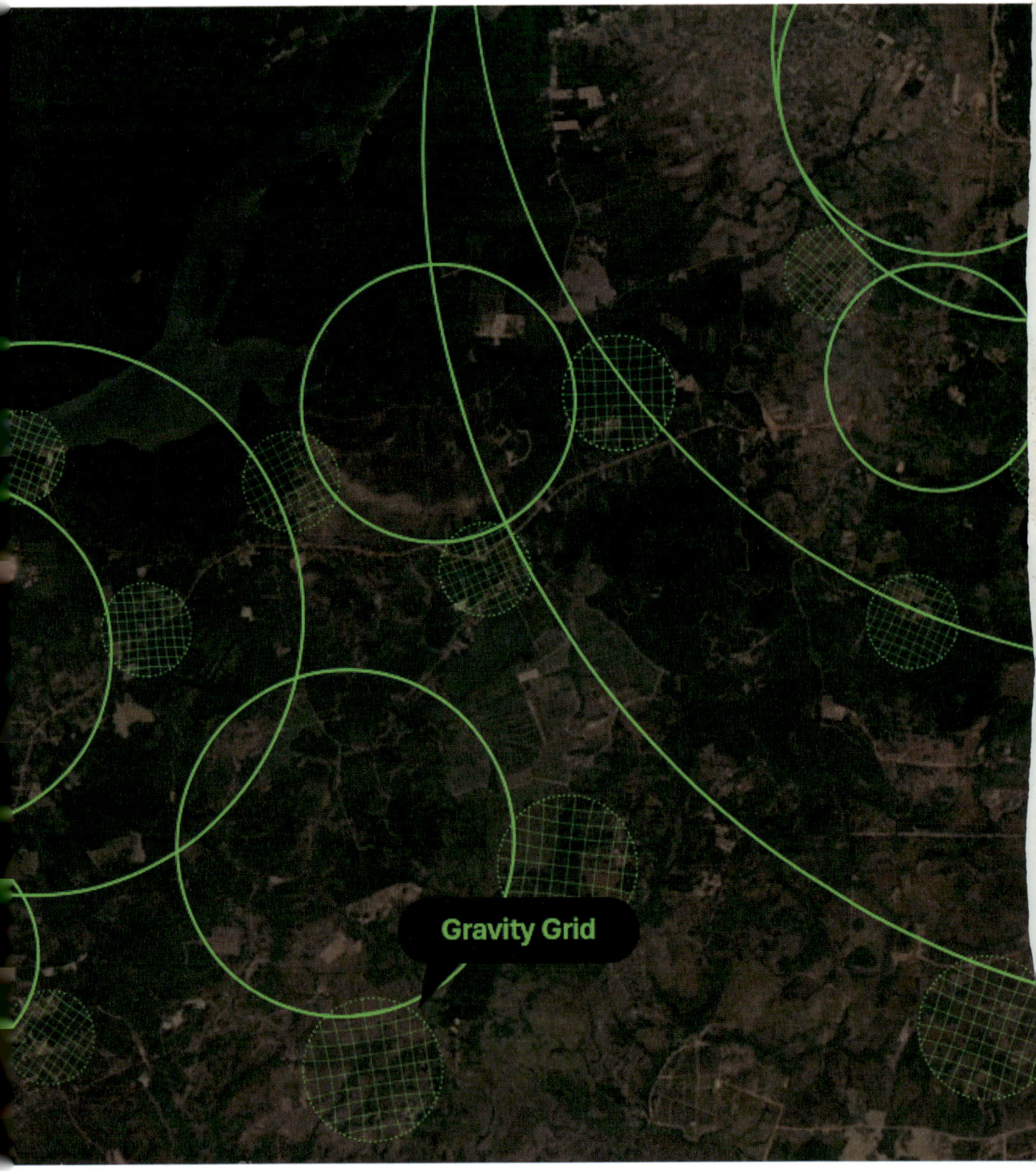
Gravity Grid

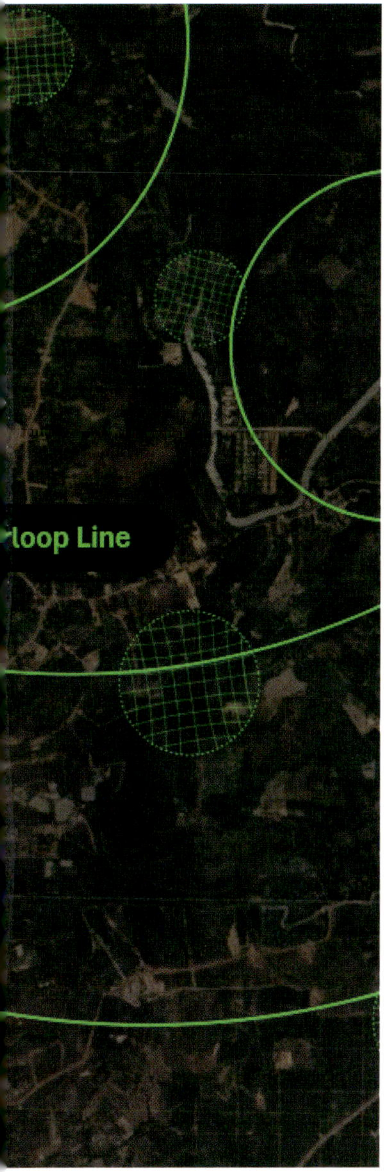

Hyperloop Connection

The Gravity Grid, which connects the city and the city, connects the city to the inside of the city. The connection between cities can be quickly moved to a hyperloop created by combining wheels.

The merged wheel can be separated into an individual inside the city and function as a PM(Personal Mobility).

Through this, the process of arriving in front of the house from the movement between cities can be seamlessly performed.

Seamless Transit system

The separated wheel moves along the Gravity Grid that runs inside the city. Using this, not only goes to the front of the house at the hyperloop station, but also moves inside the city seamlessly.

Design Principles

1. Landuse

Land use is divided into six area elements and two point elements.

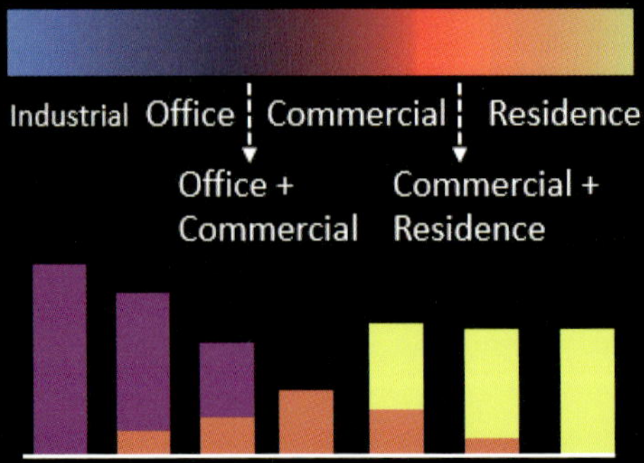

Where the two uses meet, the two uses are mixed through a vertically compounded plan.

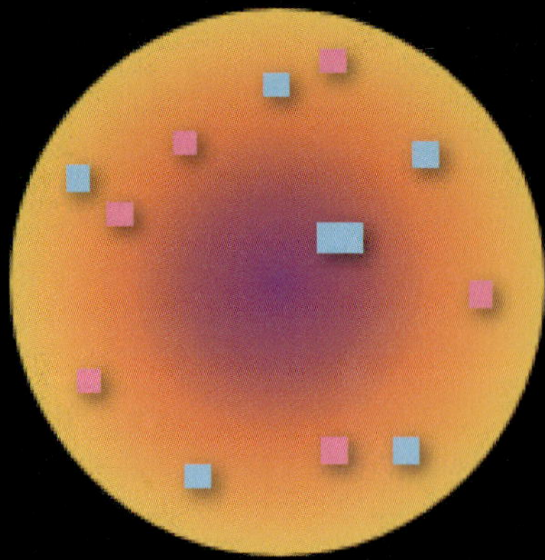

The city is made up of gradual zoning, and cultural and educational facilities are created here and there like chocolate chips according to the needs of citizens.

2. Green System

The city's main green axis flows in the form of a cross from nature outside the city.

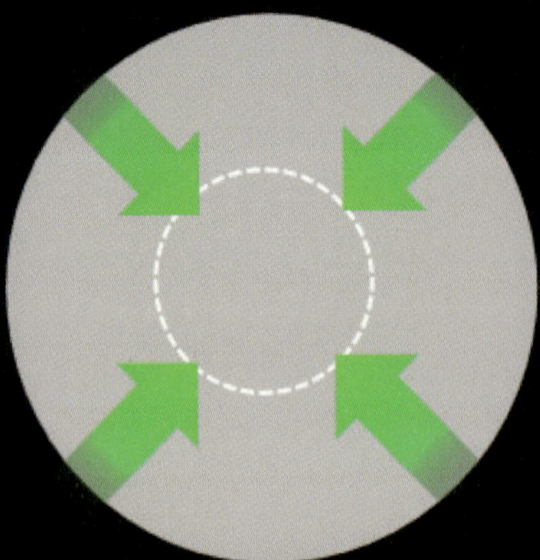

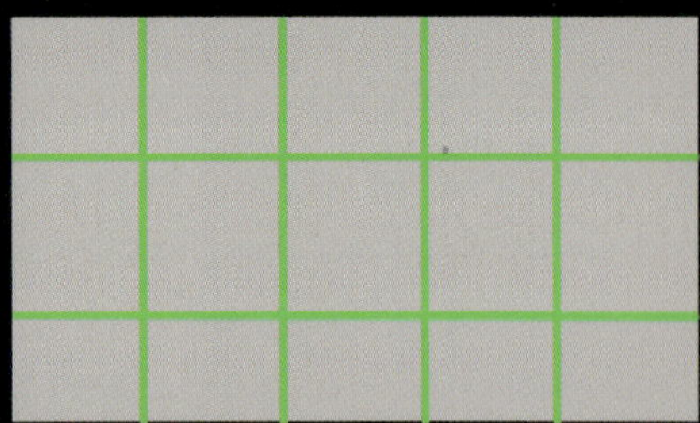

Since there are no cars in the city, the road between the blocks is green to secure green area.

Existing roads for cars require more than 5m per lane, but roads for SMwheeL are abou
1.5m per lane, and the maximum lane is two lanes.

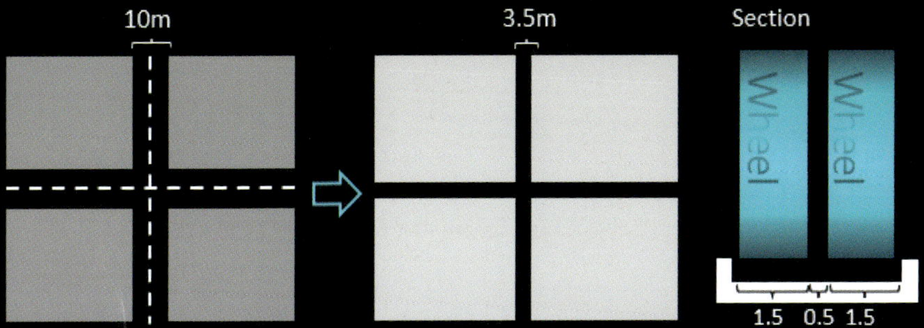

Existing road system for cars

New road system for new mobility

Wheel system

4. Block Design Principles

Existing roads for cars require more than 5m per lane, but roads for SMwheeL are abou
1.5m per lane, and the maximum lane is two lanes.

Each block has a green space in the middle of the buildings to function as a community.

On the ground floor of the block, a walking passage is opened diagonally.

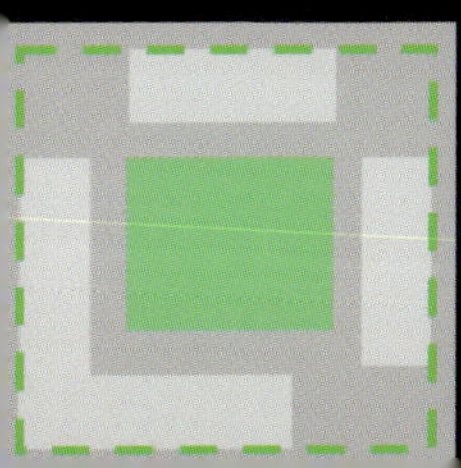

The separation distance of a building from the land boundary of the block shall be 1.5m that can be walked.

Urban Section

The city's cross-section is as follows. As the skyline shows, the city center is dense, and the more dense the building goes to the outskirts, the more dense the building is placed. This allows use and density to spread like gradation without any obvious boundaries. Contrary to the density of the building, the green space widens toward the outside, increasing carbon absorption outside the city.

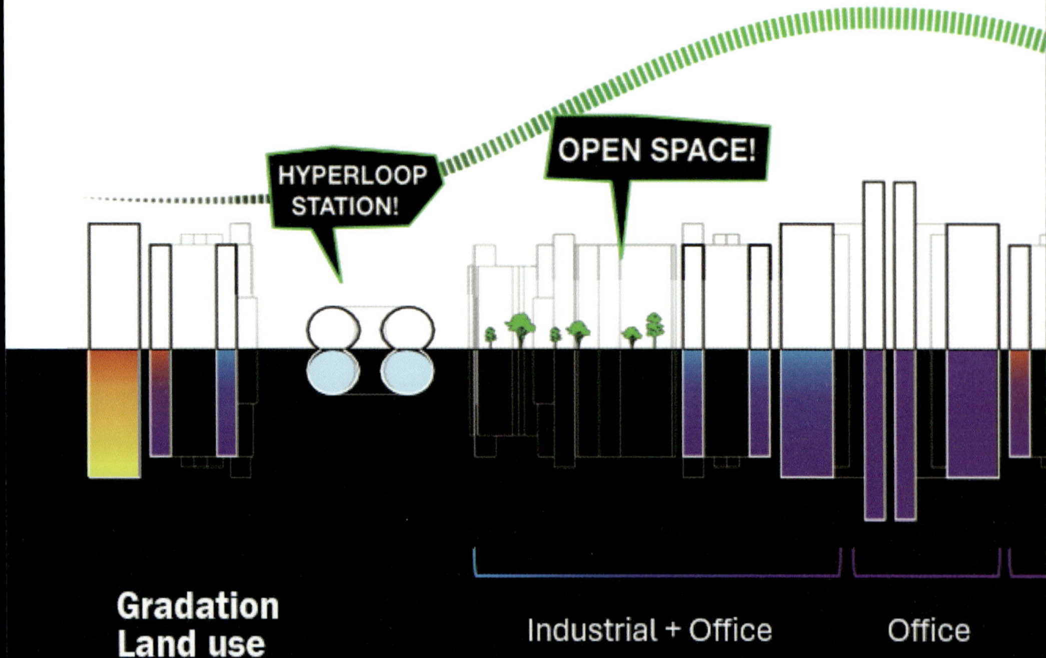

Land-use Map

The plan view of the use of the city is shown in the picture on the right. The boundary between the blocks can be broken by land use for vertical and complex uses.

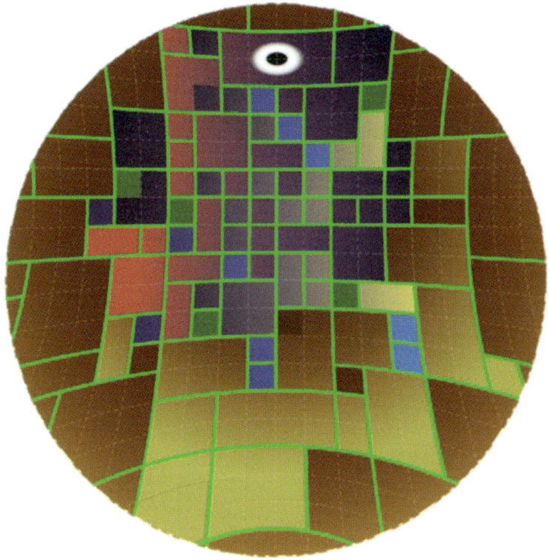

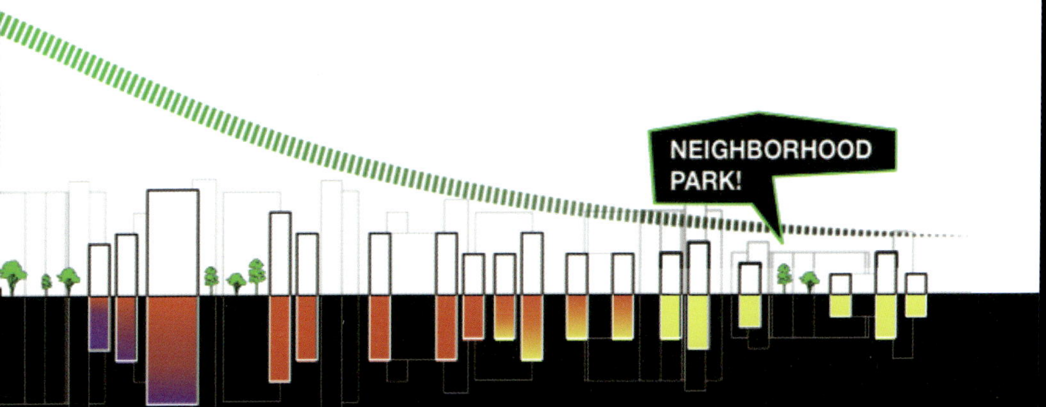

Masterplan

First of all, green areas that can adjust the size of open space to fit the block size created to fit the gravity grid were arranged. Pocket parks were created between buildings in high-density areas, and larger green areas were created in low-density areas outside the city so that they could be connected to natural green areas outside the city.

Underground Plan

There is a hyperloop station underground space, which is the base for logistics and water resources. And along the grid, the movement of logistics is automated and there is an irrigation passage. Depending on the needs of citizens, you can select and insert the warehouse module or smart farm module.

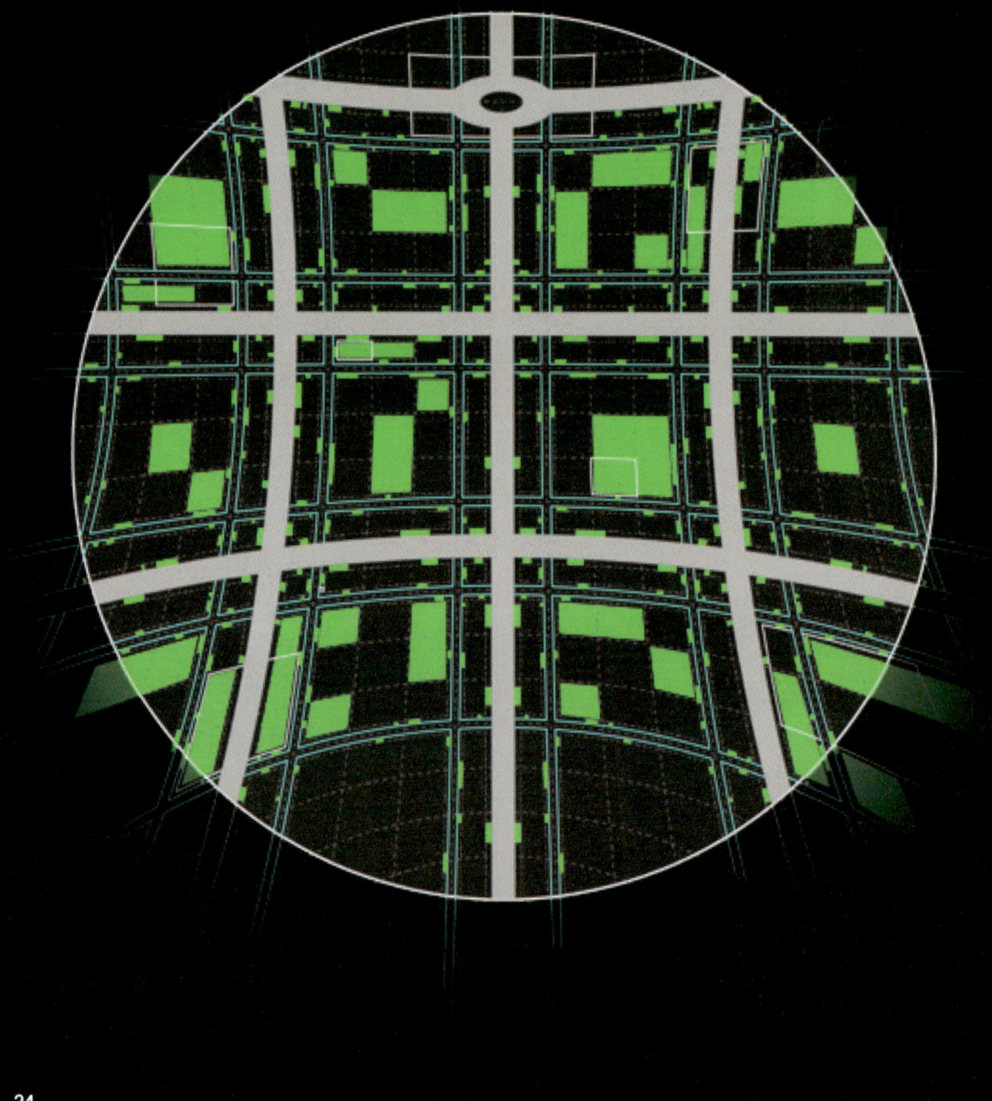

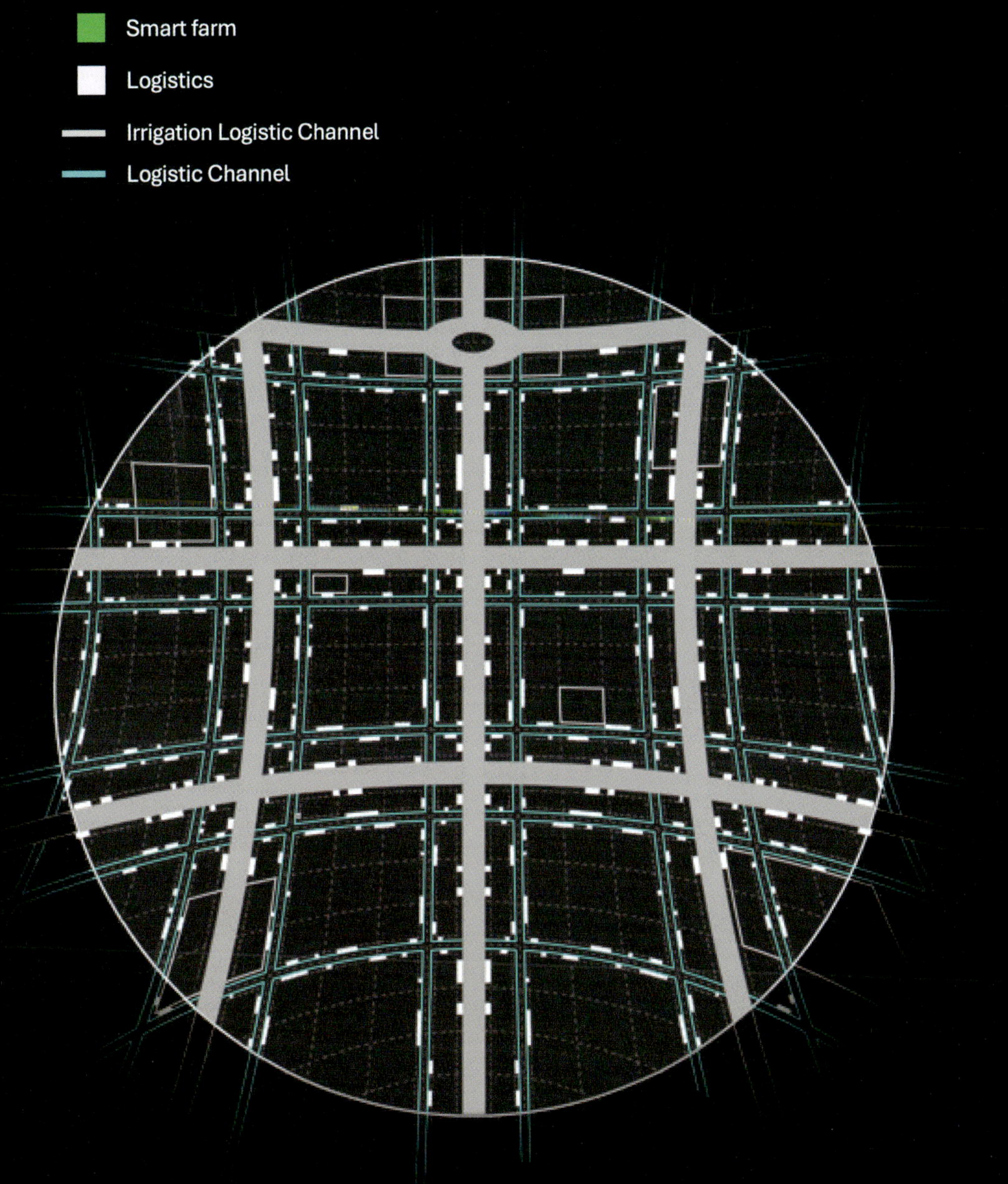
Smart farm
Logistics
Irrigation Logistic Channel
Logistic Channel

Aerial View

Hyperloop Station

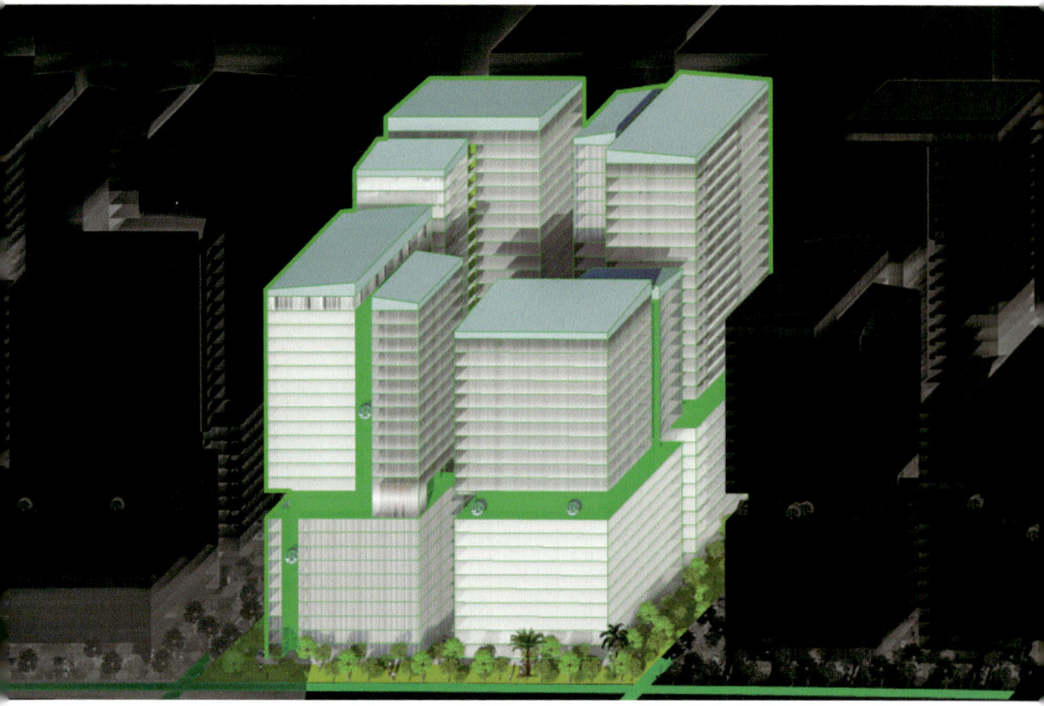

High – density Block

A high-density block is a district where a combination of industrial and business is used. Here, wheels can move along the road built on the building, allowing the ground to be more yield to pedestrians. Industrial programs are arranged on the floor where the wheel roads are planned to facilitate the movement of logistics. The logistics tube created underground extends to the interior of the building, thereby maximizing the efficiency of the city as a whole.

Wheel roads inside the building

Mid – density Block

Medium-sized blocks are used for commercial and business purposes. There is a smart farm on the top floor of the building, which can contribute to the self-sufficiency of the city. In the center of the block, a park surrounded by buildings was created to promote the connection between the green areas inside and outside the city. The park was planned as an open green area that gives rest and space to nearby residents and users of offices and commercial facilities in medium-sized blocks.

The park inside mid-density block

Low – density Block

In the low-density block, several residential units are placed on the right side of the corridor, and a large urban forest is created among them. The lower stratum of the corridor is an open space, and serves as a community space where various events can take place. Through this open space, residents can use the urban forest without going through the residential area. The wheel extends to the inside of the urban forest so that you can conveniently use the flow across the forest.

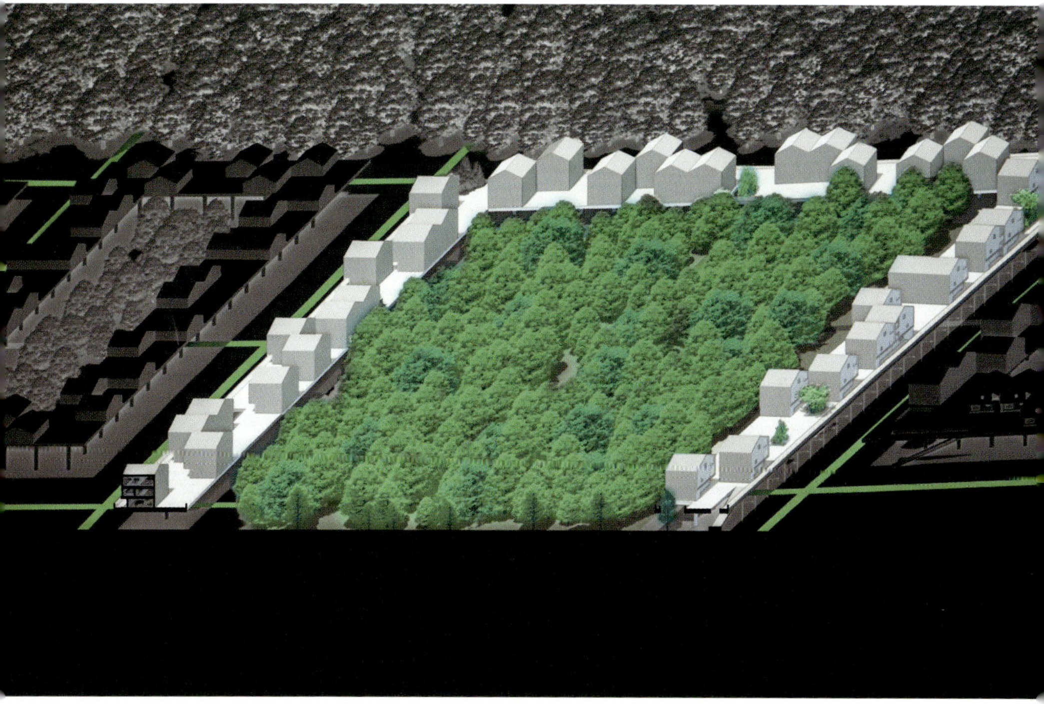

Low-density block from the user's point of view

Circular Zero City

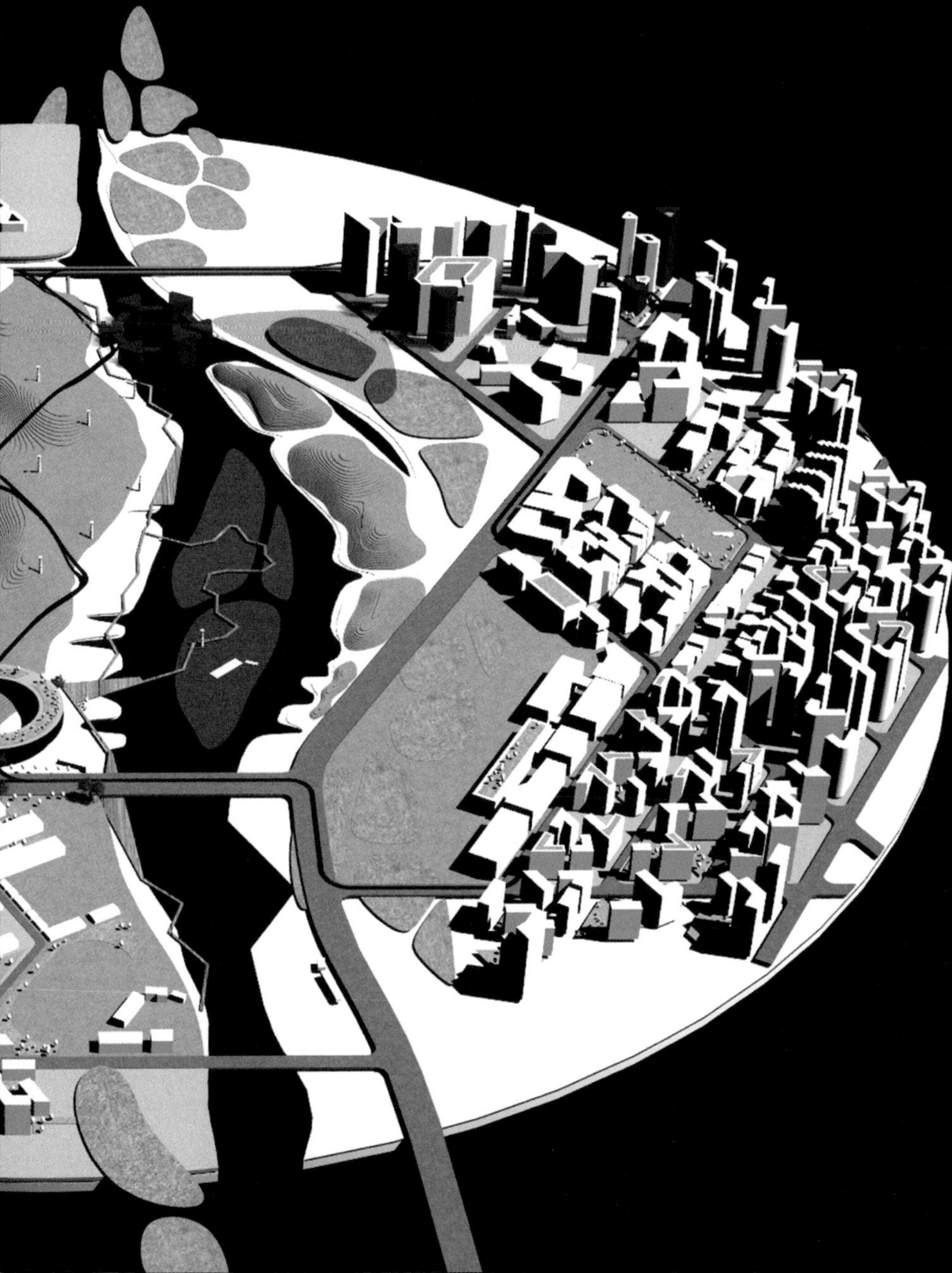

Circular Zero City

2조 : 임영빈, 조우진, 차하림

Challenges in 2050: The Problems facing Future Cities

In 2050, the metropolitan cities made remarkable progress but faced numerous problems at the same time. Cities filled with skyscrapers and high-tech technologies are thriving on the outside, but complex and deep problems are hidden behind them.

First, the problem is overcrowding. As countless people flock to cities in search of better lives, housing shortages and traffic congestion worsen, and the gap between the rich and the poor widens. It is difficult to find a clue to the problem of slums and homelessness in urban areas. Citizens live by day to day, and urban infrastructure is reaching its limit.

Second, the environmental destruction is serious. Due to reckless development and overuse of resources, urban green spaces are gradually shrinking, and air and water pollution have become a daily problem. People struggle to find clean air and water, and urban ecosystems are gradually destroyed.

Third, the frequency of natural disasters caused by climate change increases. As extreme climate phenomena such as heat waves, floods, and typhoons become more frequent, urban infrastructure is greatly threatened. Climate change is no longer a story of the distant future and is currently taking a direct hit on city life.

To solve these problems, innovative approaches and sustainable policies are needed. Various measures such as population distribution policies, eco-friendly city development, and climate change response strategies must be sought. The challenges of 2050 are only the beginning. The future of the city is in our hands.

It is not enough to solve the problem only with large cities. A new solution was needed. It is the introduction of the "Compact city."

Compact city have emerged as a new paradigm for solving problems in large cities. The compact city, self-sufficient cities based on cutting-edge technology and sustainability are the solutions. It is a new type of city where you can maximize the effect of population distribution and live in harmony with nature through environmentally friendly urban design.

The core of the compact city is distributed resources and energy management systems. It has a system that can generate its own energy, recycle waste, and become self-sufficient in food. This is very helpful in mitigating the overcrowding of large cities and providing people with a more pleasant living environment.

In addition, the compact city is a people-centered city. With the application of advanced smart city functions, residents can enjoy safer and more convenient lives, and the concentration of quality-of-life factors such as education, medical care, and cultural facilities in close proximity makes it easy for them to use all necessary services without having to travel far. In addition, it is effective in responding to climate change that we are facing. Through eco-friendly buildings and renewable energy systems, we will be able to minimize carbon emissions and better prepare for natural disasters caused by climate change.

In 2050, the problems facing cities in the future are complex tasks that are difficult to solve only with the efforts of large cities. The introduction and spread of compact cities can be an important key to solving these problems.

The new urban model in which large cities and small cities develop together in harmony will contribute to overcoming the challenges facing humanity and creating a sustainable future.

OpenAI. (2024). ChatGPT (ver4.0). https://chat.openai.com

Now, the challenges of 2050 are transforming into new possibilities through the compact city. The future of cities is in our hands, and only smart choices and innovative approaches will provide the solutions.

CIRCULAR ZERO CITY: Concept

Circular Zero City is not just a city, but a special place designed based on learning, nature, and human connection.

The campuses distributed throughout the city offer learning opportunities anywhere, so small classrooms appear around the corners of the streets, and classes can be taken freely on park benches. People explored academics, shared knowledge, and realized new ideas at their own pace and way.

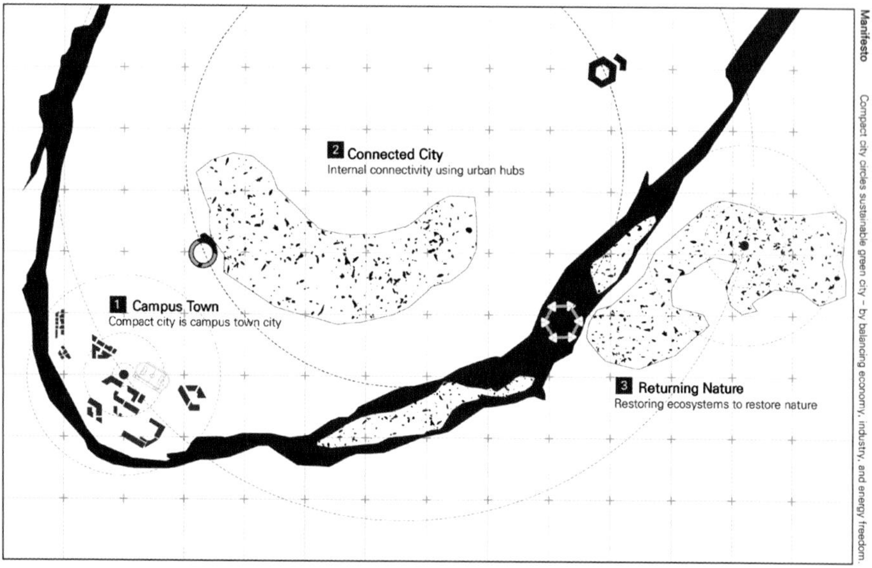

In the center of the city, there is also a vast expanse of green space. People seek peace of mind in contact with nature, free from daily stress. Parks, gardens, and trails spanning the city make it easy for anyone to feel the nature.

The core of the city is that different hubs are organically connected. These hubs will make the lives of urban residents more convenient and create a sustainable environment for future generations.

The first hub is the Green Hub. It cleans contaminated land, studies animals and plants in the green space, and manages the entire green space network. The Green Hub is a landmark and cultural space facility located in the center of the city, and plays an important role in connecting people and nature.

The second hub is the Data Hub. It is a center that collects and manages all the city's data. It systematically manages a variety of data from city infrastructure, medical care, green areas, and plants, thereby maximizing city efficiency and leading residents' lives in a better direction.

Lastly, the Medical Hub. It is a medical facility responsible for the health of its residents, which solves both large and small diseases, and is always by its residents. The Medical Hub combines cutting-edge medical technology with human touch to create an environment where residents can live with confidence. The city's residents live their own lives, connect with each other through the hubs, and realize learning in nature.

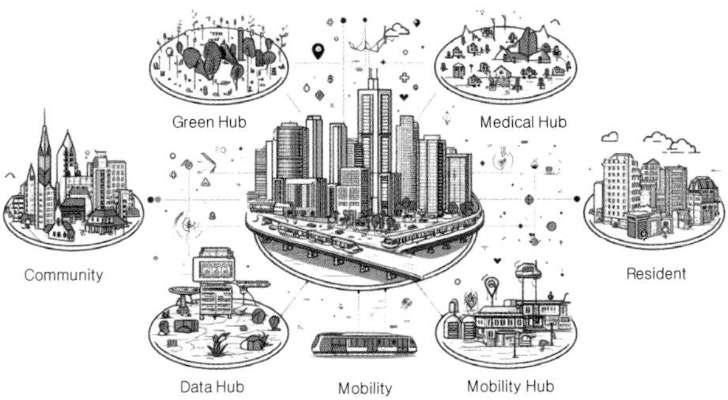

OpenAI. (2024). ChatGPT (ver4.0). https://chat.openai.com

CIRCULAR ZERO CITY: Urban Greening

We will greenize 80% of the city and return nature destroyed by climate change and environmental pollution to forests, grasses, and rivers.

Cities are reborn as forests, wetlands, and parks through ecological restoration of contaminated land, and can face green forests and rivers anywhere.

The work of blocking pollutants through the soil cover method and planting tree species that decompose pollutants and returning them to nature is carried out. Urban forests and natural rivers are created so that residents can enjoy rest and healing in a clean environment.

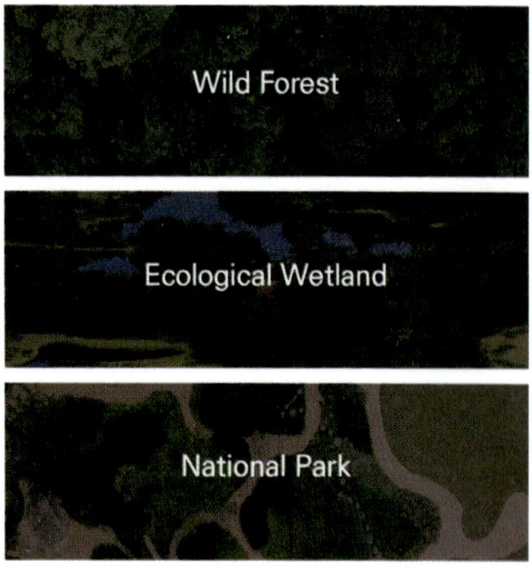

The utilization of green space

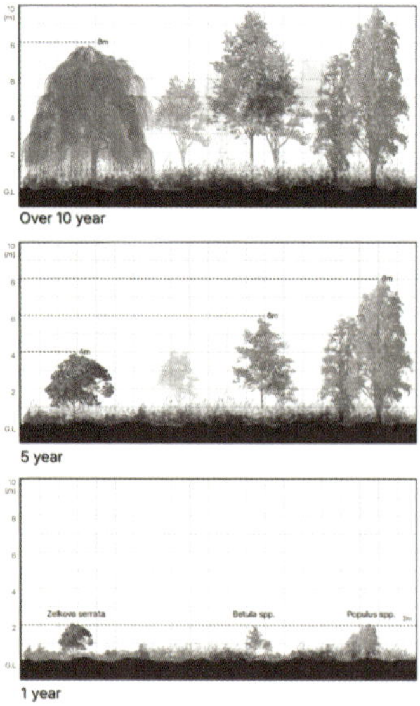

The project uses fast-growing tree species to quickly form forests.

Pipulus spp.(1.5~2.5m) > Betula spp.(0.5~1m) > Zelkova serrata(0.3 to 0.5m annual growth)

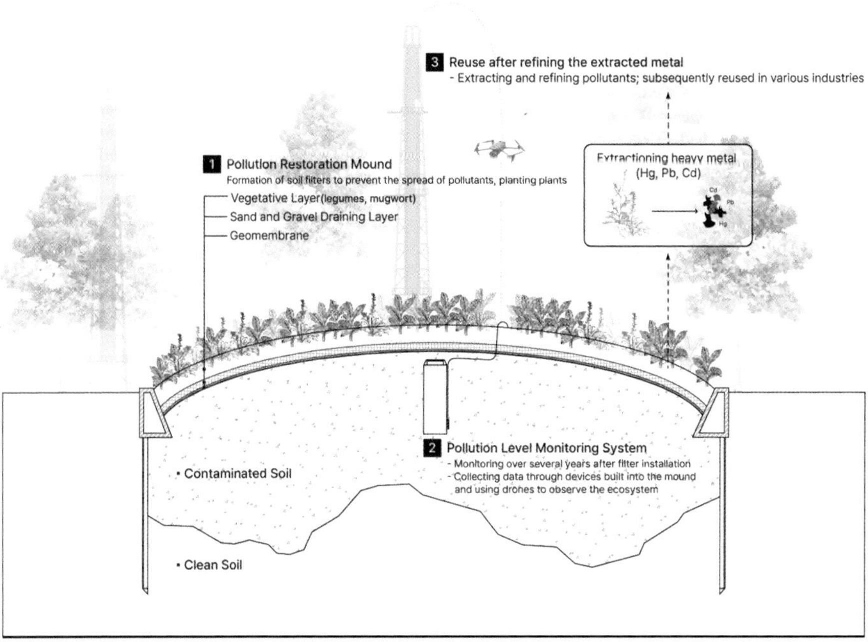

Install impermeable plates to monitor heavy metal pollutants in the depths of the land
Mound has a minimum length of 20m and a maximum length of 50m

CIRCULAR ZERO CITY: Triangular Grid Innovation

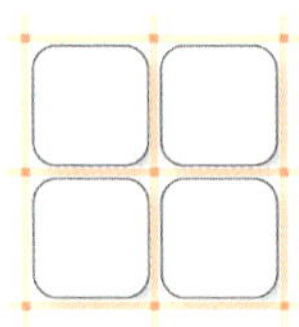

Rectangular Grid

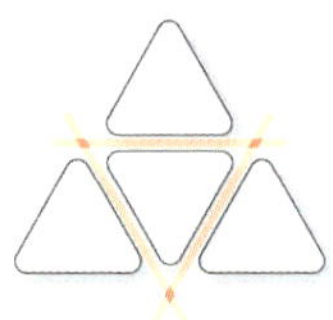

Triangular Grid

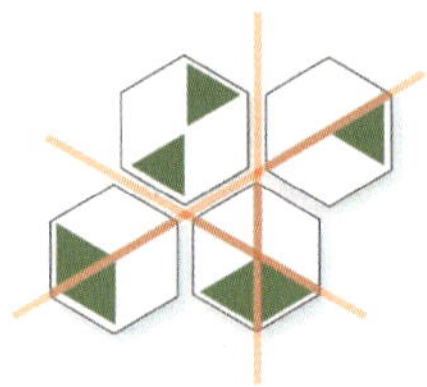

Hexagon Grid

We have introduced a unique triangular grid for efficient city operation. Access and connectivity between the blocks that make up the city have been maximized.

In the triangular grid, each block is not independent but organically connected to each other, so the entire city works like a living thing. We formed various block groups around each node by constructing blocks based on the triangular grid. This becomes an important factor in helping each part of the city to function in harmony.

Triangular grids have many advantages in terms of efficiency, economy, and adaptability. Compared to existing rectangular grids, triangular grids can reduce road area, thereby mitigating traffic congestion and maximizing space utilization in cities. It can be easily applied to irregular terrain, minimizing land waste, which will greatly reduce the cost of land use and reducing the economic burden on cities.

Circular Zero City further increases the efficiency of the city through a triangular grid. The grid allows for fast connections throughout the city and efficient use of all space. This allowed the city to transform into a smarter, more economical, and more adaptive space.

The introduction of the triangular grid is not just a change in block composition. It is an important step for a sustainable future pursued by Circular Zero City. In the future, all spaces in our city are organically connected, enabling a new type of urban life where nature and humans coexist harmoniously.

CIRCULAR ZERO CITY: Green Network

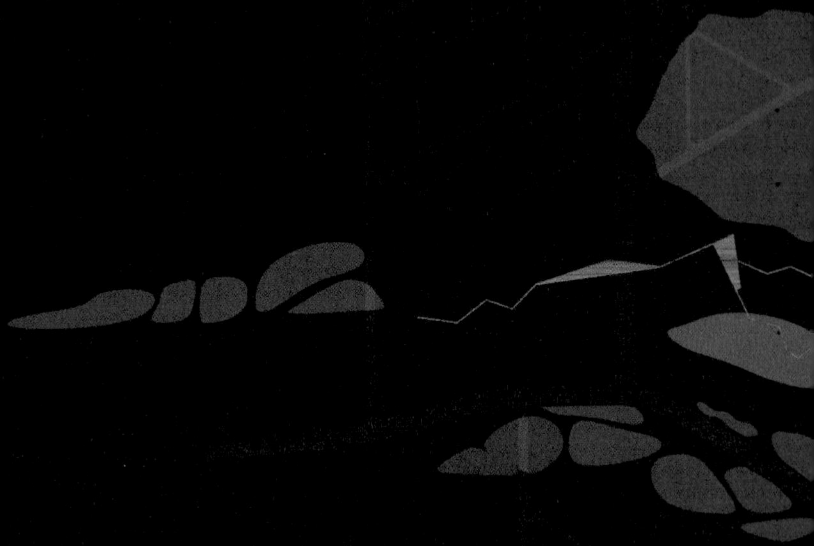

Green Area (past contaminated land)

Pedestrian Deck

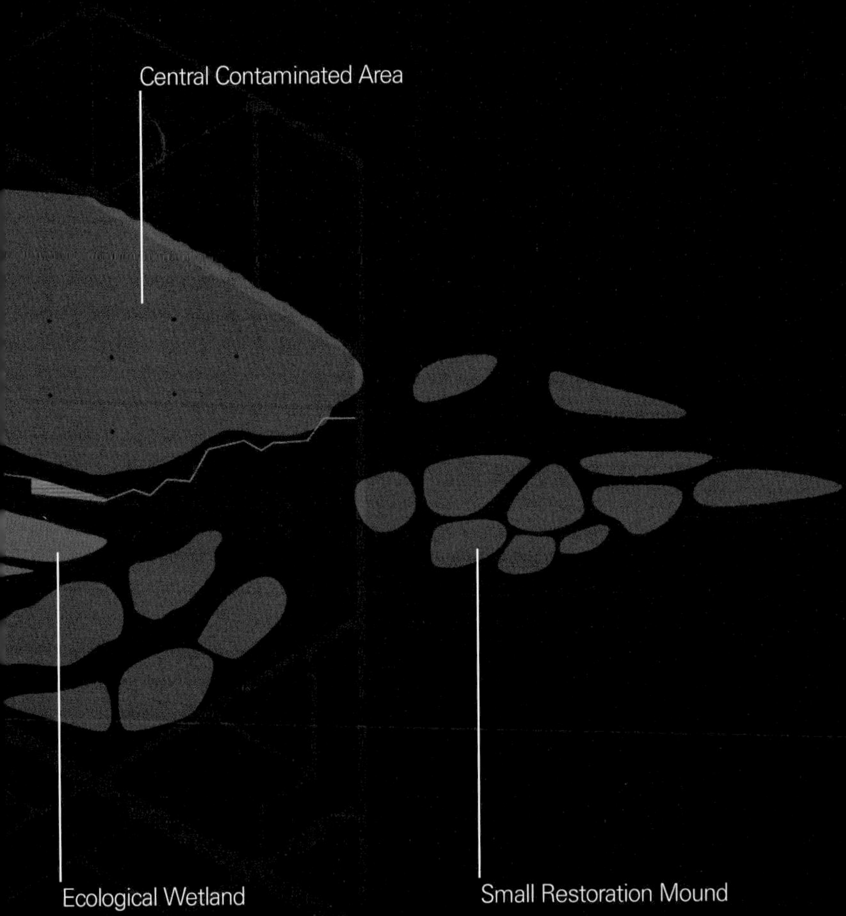
Central Contaminated Area
Ecological Wetland
Small Restoration Mound

CIRCULAR ZERO CITY: Mobility

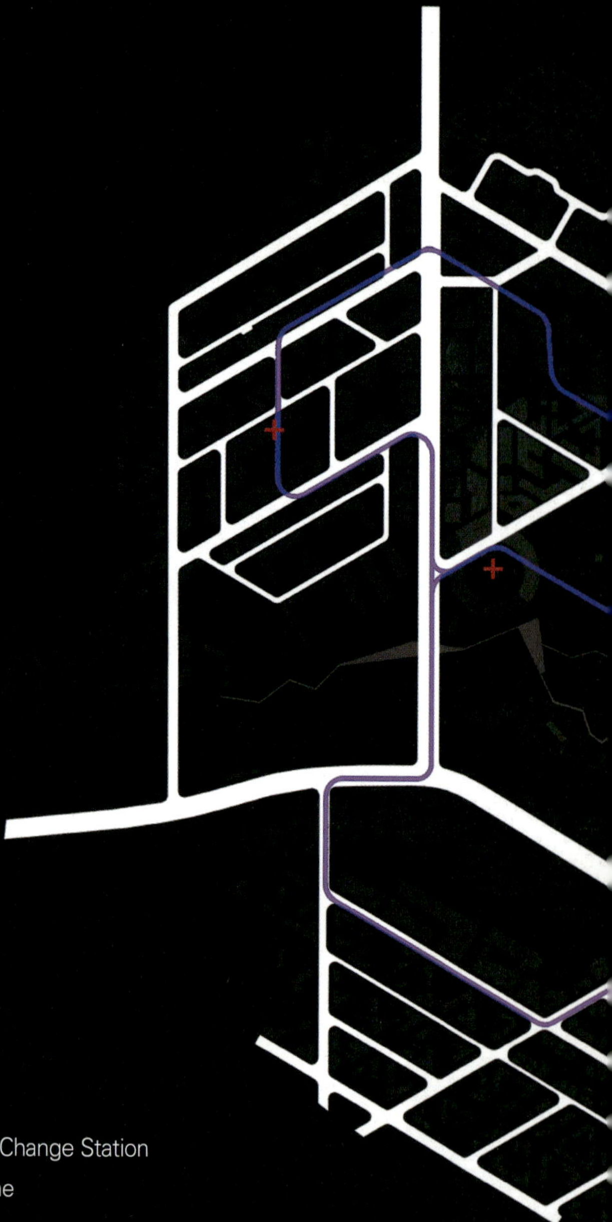

Tram – Gondola Change Station

Flexible Tram Line

Road Network

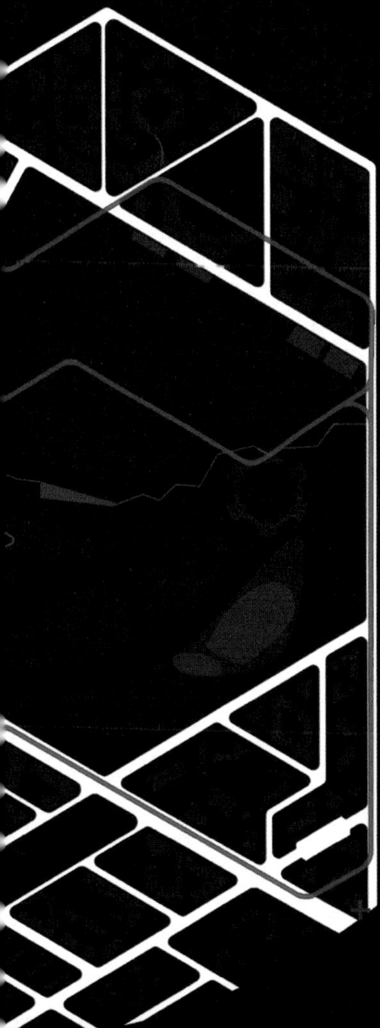

CIRCULAR ZERO CITY: Hub & Campus

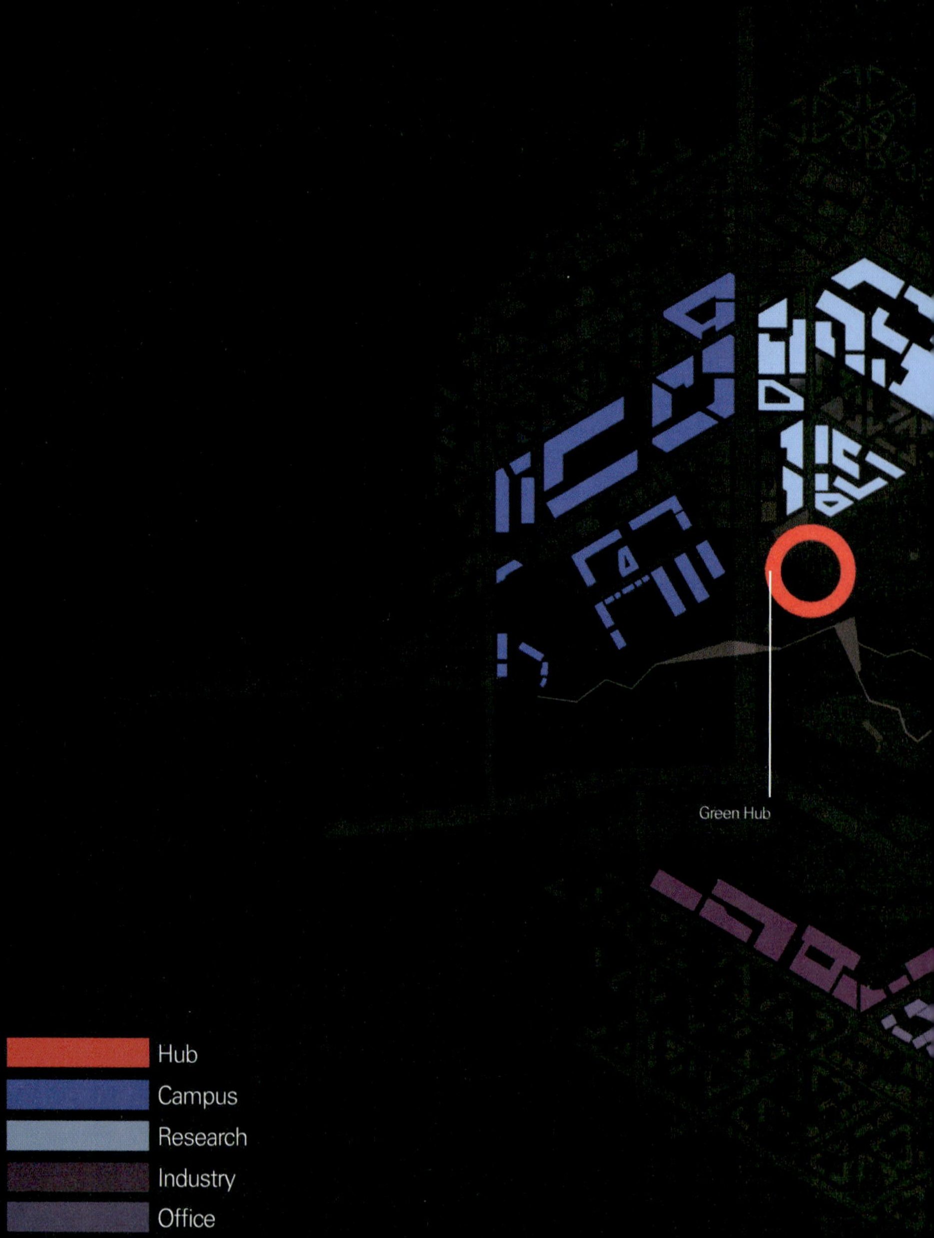

Medical Hub
Mobility Hub
Data Hub

CIRCULAR ZERO CITY: Residence & Community

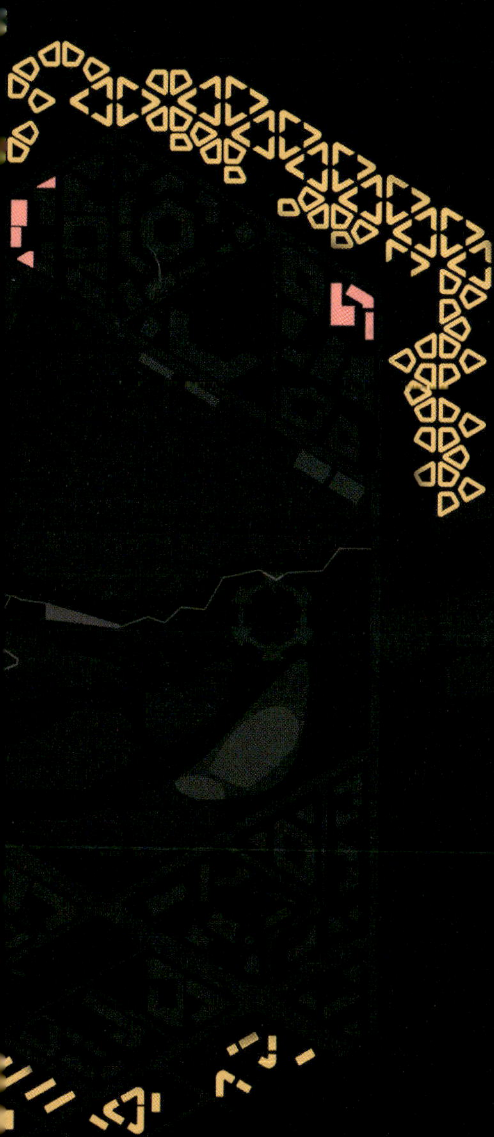

Mast

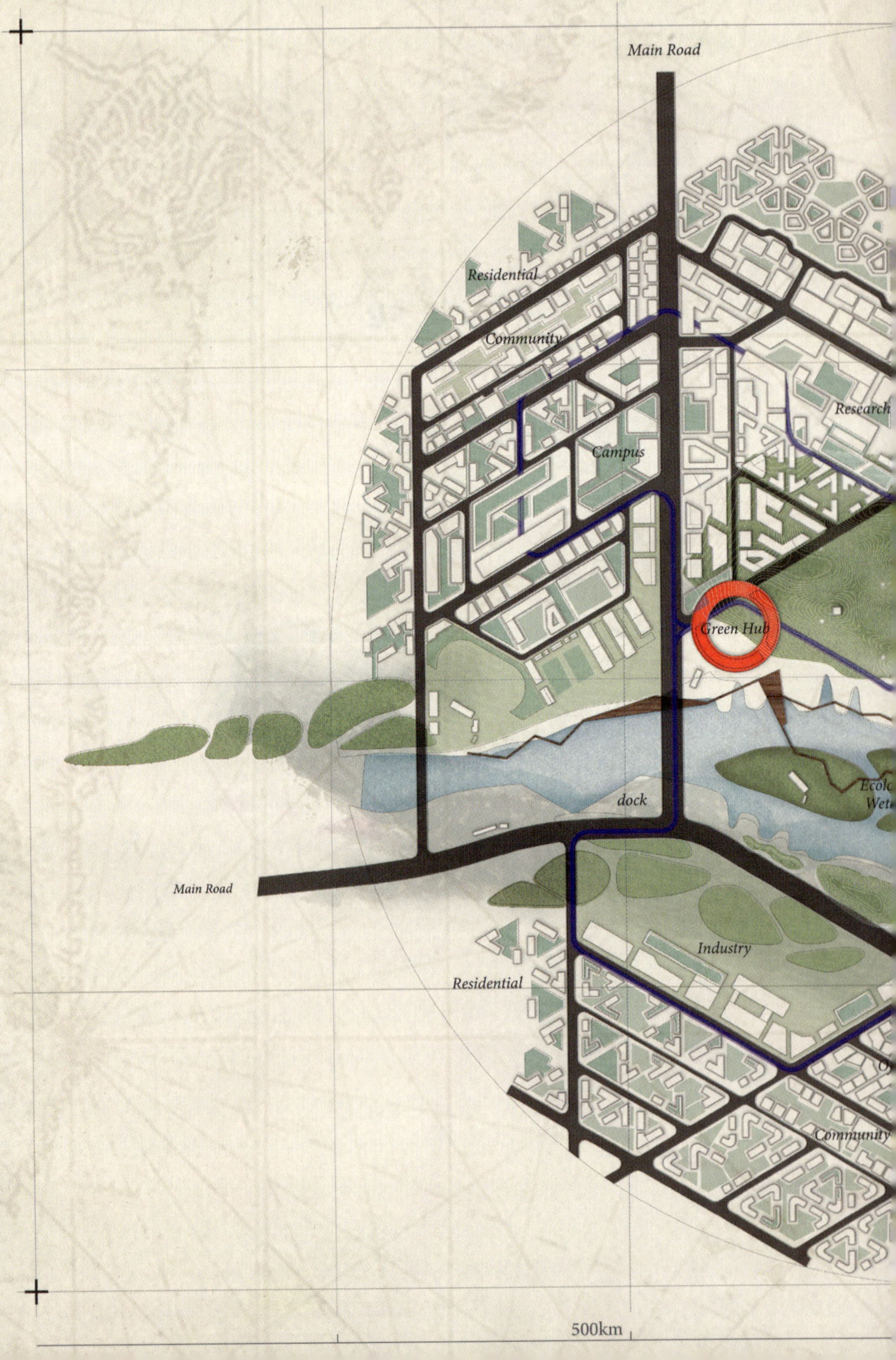
Main Road
Residential
Community
Research
Campus
Green Hub
dock
Main Road
Industry
Residential
Community
500km

Plan

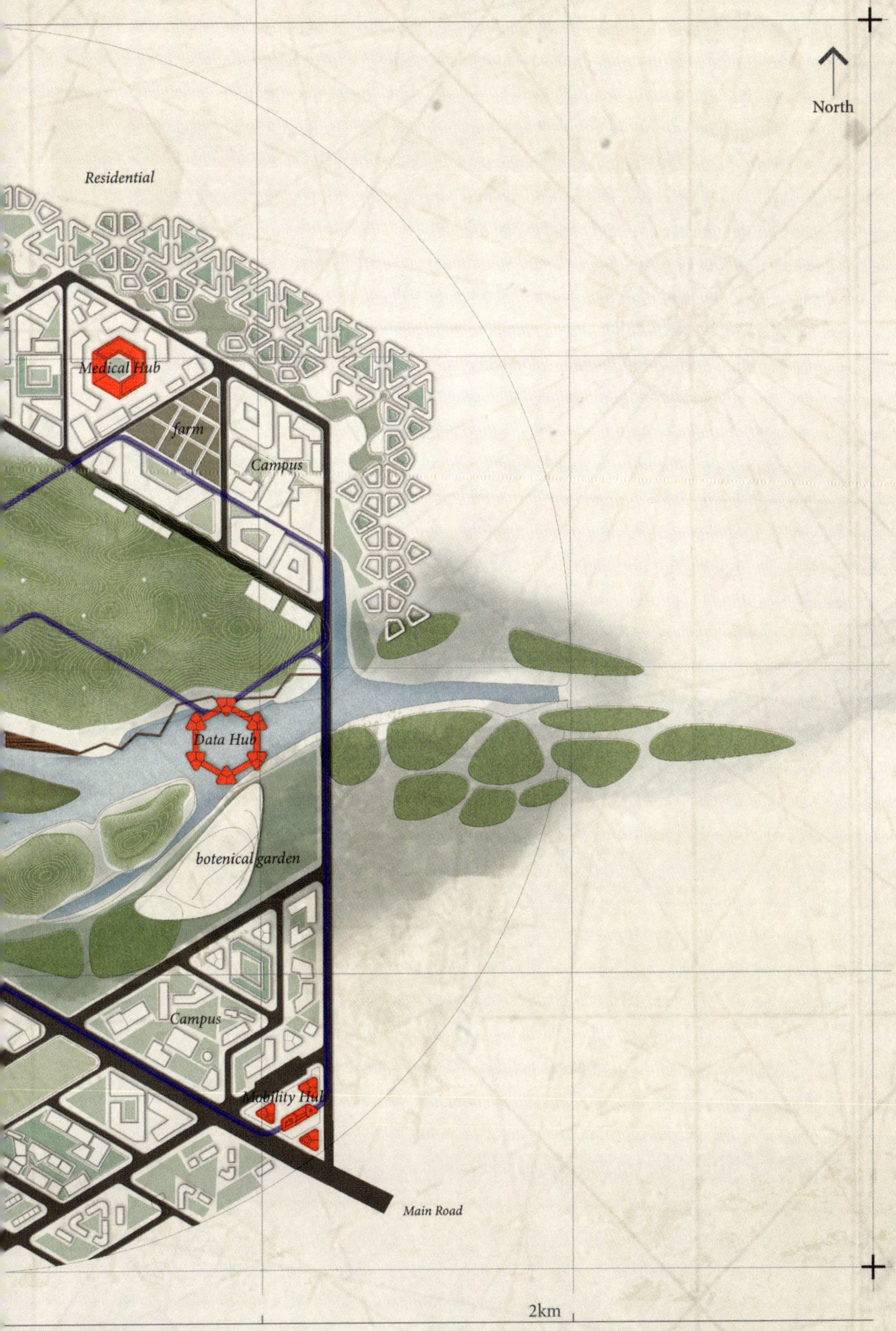
North
Residential
Medical Hub
farm
Campus
Data Hub
botenical garden
Campus
Mobility Hub
Main Road
2km

CIRCULAR ZERO CITY: Green Network

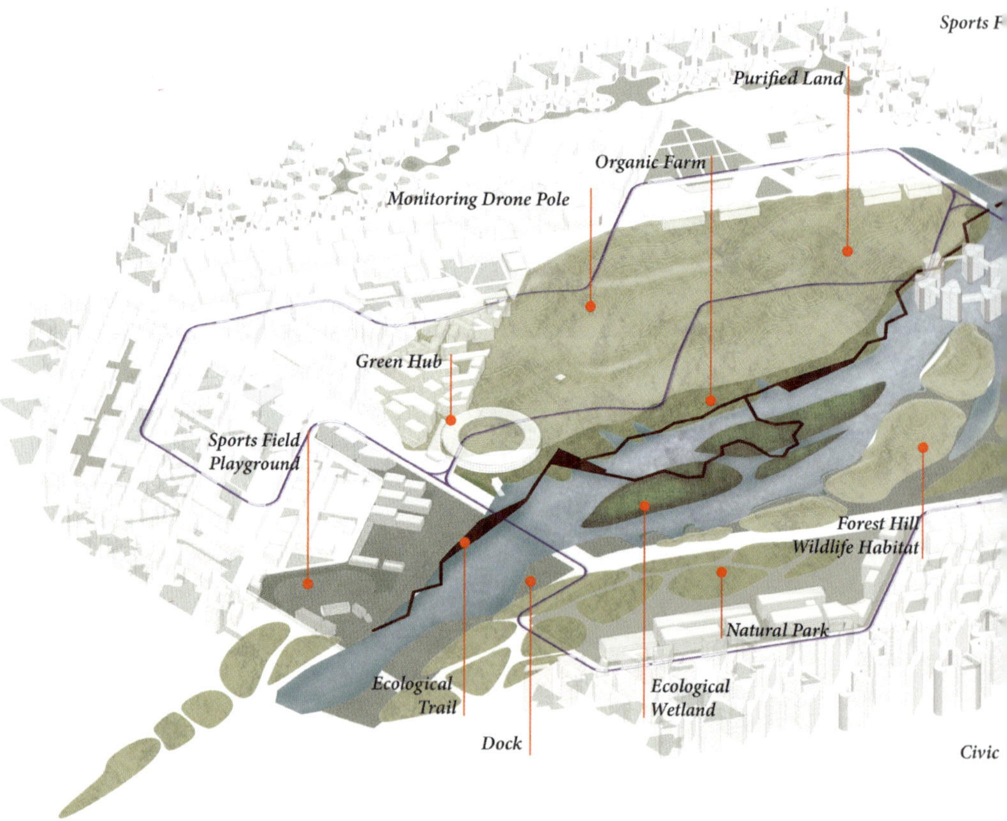

수계 | 자연녹지

하천(25%) 533,800 | 생태습지(10%) 314,000 | 야생 숲(20%) 628,000 | 오염복원녹지(15%) 471,

Circular Zero City is a city where you can meet green nature anywhere. As shown in the picture, we realized a city that lives with nature by minimizing artificial structures.

This city is not just about bringing nature into the city, but a space where the city itself coexists in harmony with nature. Parks, forests, and rivers are spread throughout the city, so residents can relax and recharge in nature at any time.

Circular Zero City leads a sustainable future through a truly green network, enabling a new form of urban life where nature and humans coexist harmoniously.

Residential Complex

Residential complex and community facilities in Circular Zero City are organically connected. All residential facilities are arranged in consideration of the skyline and the right to sunlight, and as community facilities, they are designed so that residents can enjoy various programs.

Each residential facility is closely related to the hub located around it, and the density and number of floors are planned according to its nature.

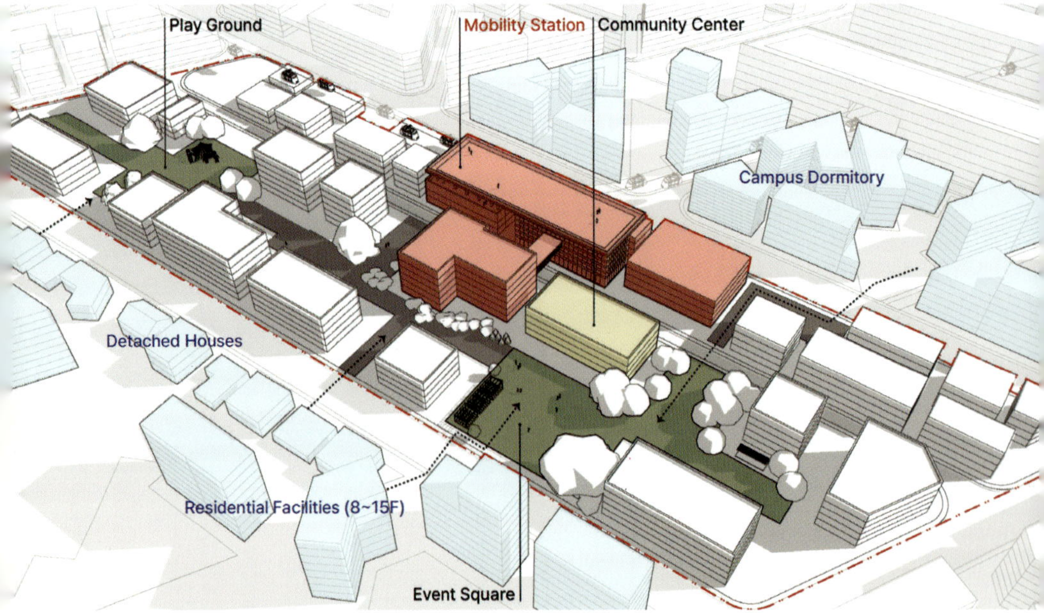

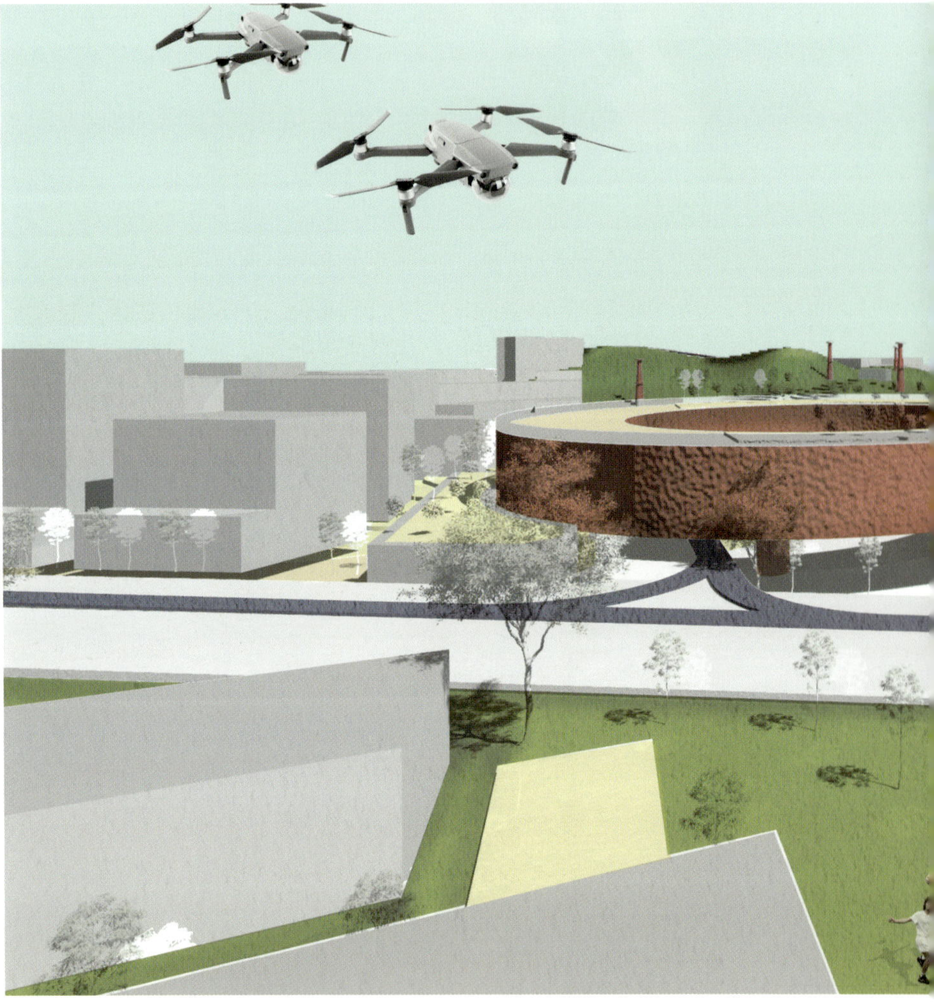

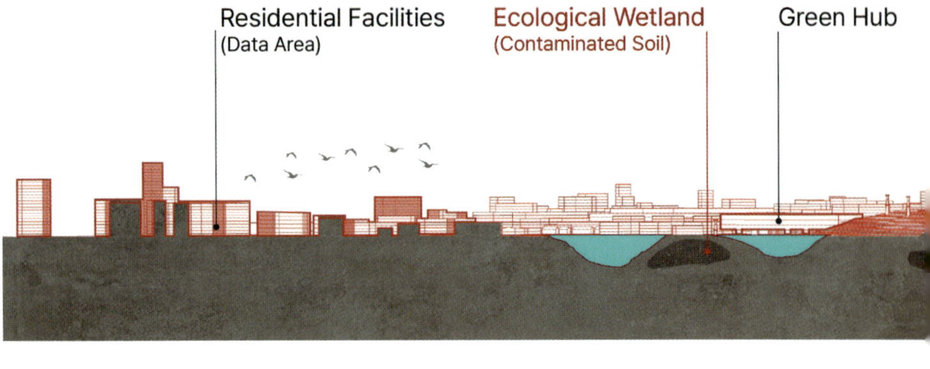
Residential Facilities
(Data Area)
Ecological Wetland
(Contaminated Soil)
Green Hub

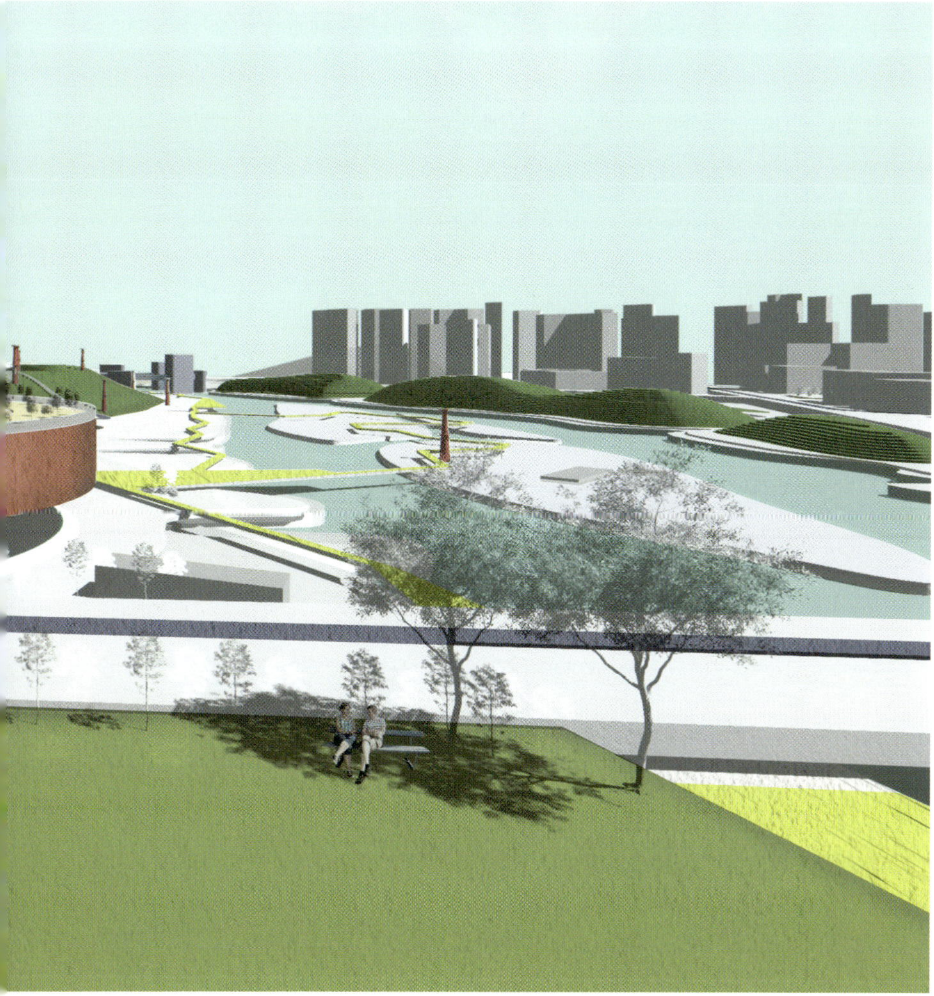

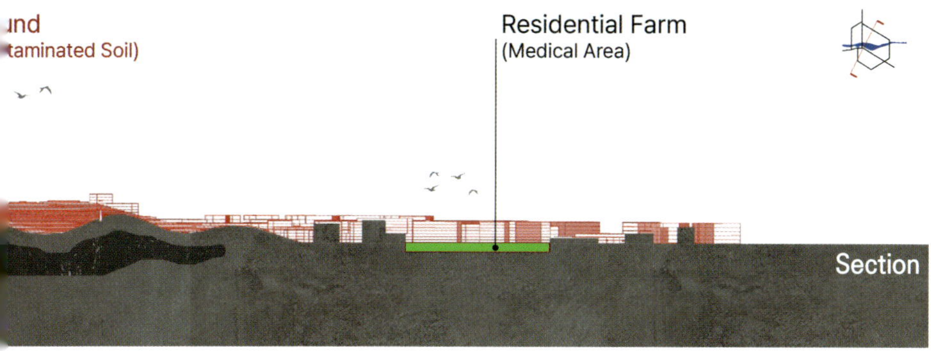
und
taminated Soil)
Residential Farm
(Medical Area)
Section

Mediacal Campus
Mound
(Contaminated

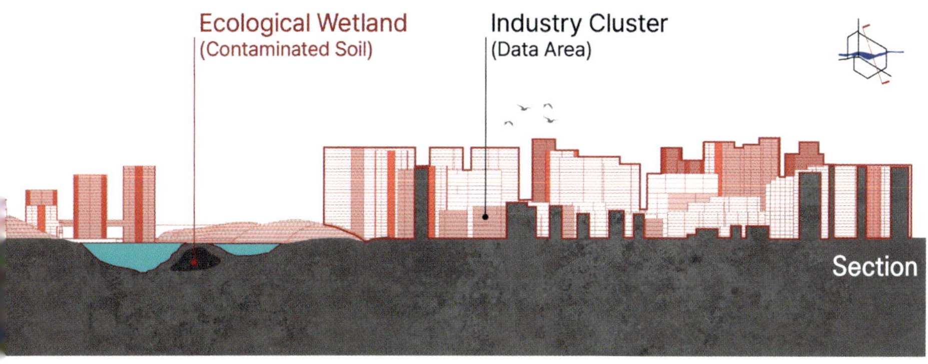
Ecological Wetland
(Contaminated Soil)
Industry Cluster
(Data Area)
Section

Mobility Hub

Circular Zero City is a city that lives in harmony with nature. It dreams of a sustainable future with the harmony of nature and humans. Accompanied by people, nature, and future generations, "**Circular Zero City**" will lead the future.

Negotiation CiTy

Negotiation CiTy

3조 : 박소연 전형욱 조인서

We imagined 'Negotiation City' which can constantly changing, flexibly negotiating and accepting the environment and the needs of city members that are changing every moment.

In the future, it will be possible to create a abundant city through 3D-based urban planning, which is no longer 2D-based urban planning. In addition, Nego city respect individual lifestyles and diversity of life that will become more diverse, but living together, and look forward to the possibility of a city that will develop more and more in the future.

This city is a planned city on a basic grid (600m x 600m) consisting of 50m x 50m block units and a 10m-wide walkway. The city is based on 3D ZONING, not 2D, and consists of a module system. Zoning boxes with different characteristics for each program are 10m x 10m units, and modules are 5m x 5m units. Programs can be negated and shared among users in the process of zoning. Modules completed after the nego may be connected to public green spaces walking routes and mobility roads.

Background

〈Problem of 2D Zoning〉

Various urban planning proposals exist to date, but most are based on 2D-based planar zoning.

〈Constant changes for ideal city〉

After the city is formed, people continue to develop new needs and to accommodate them, the city needs constant change.

Concept

〈Physical Concept〉

The city we think of as a Robust Compact City is **Tetris.**

〈Systematic Concept〉

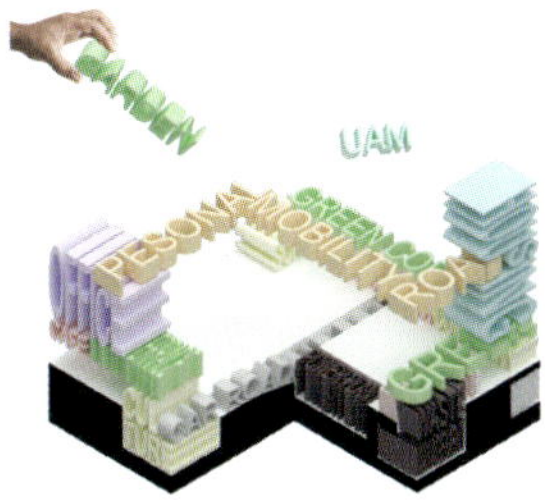

The city we think of as a Robust Compact City is **Negotiation.**

'Tetris' only exists in the x and y axes.
'Cube' can achieve three-dimensional and enables the composition of robust compact cities.

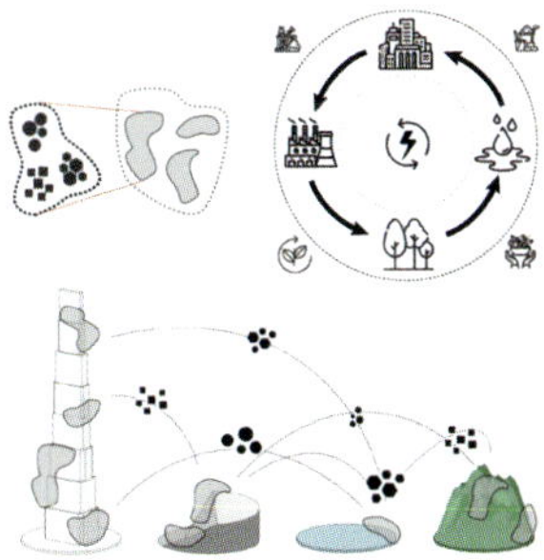

Regardless of the status of the site, various modules can be applied flexibly.

Manifesto

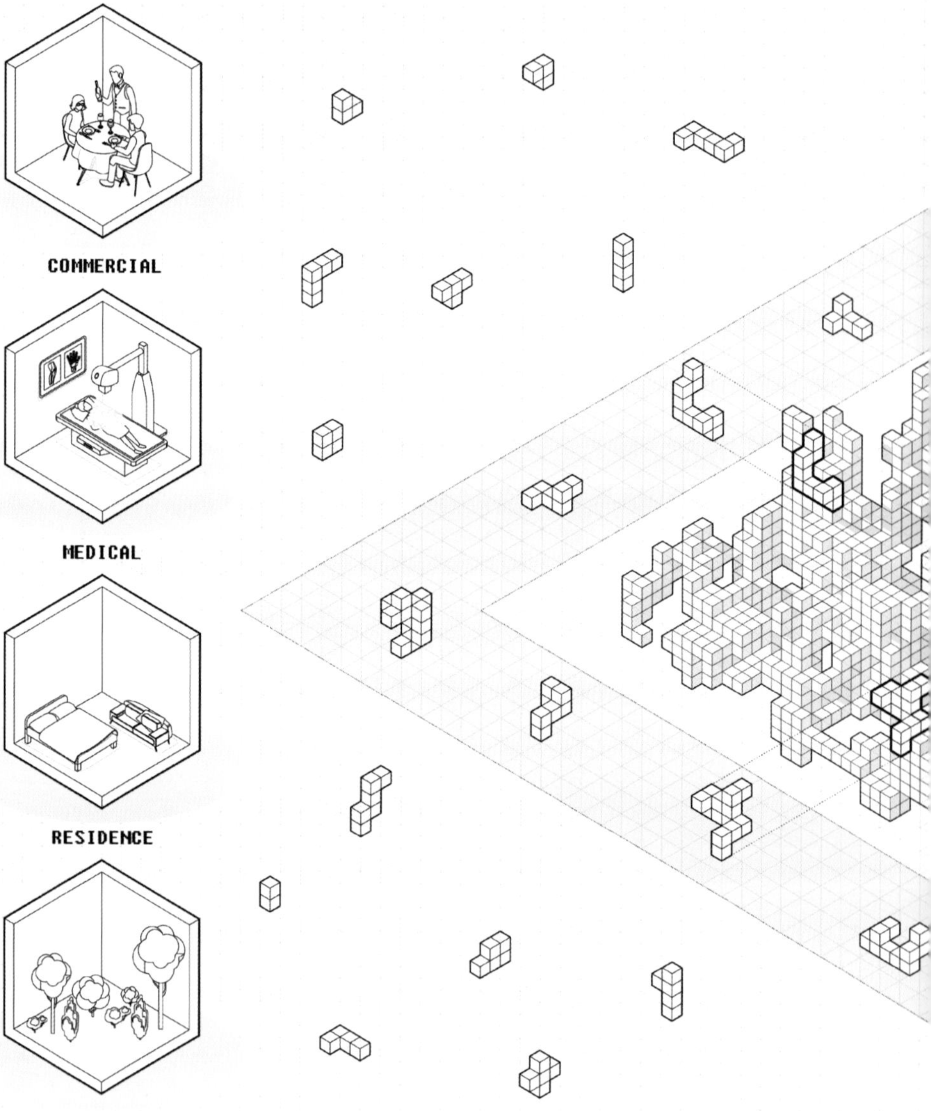

The city we think of as a Robust Compact city is ...

" Negotiation City "

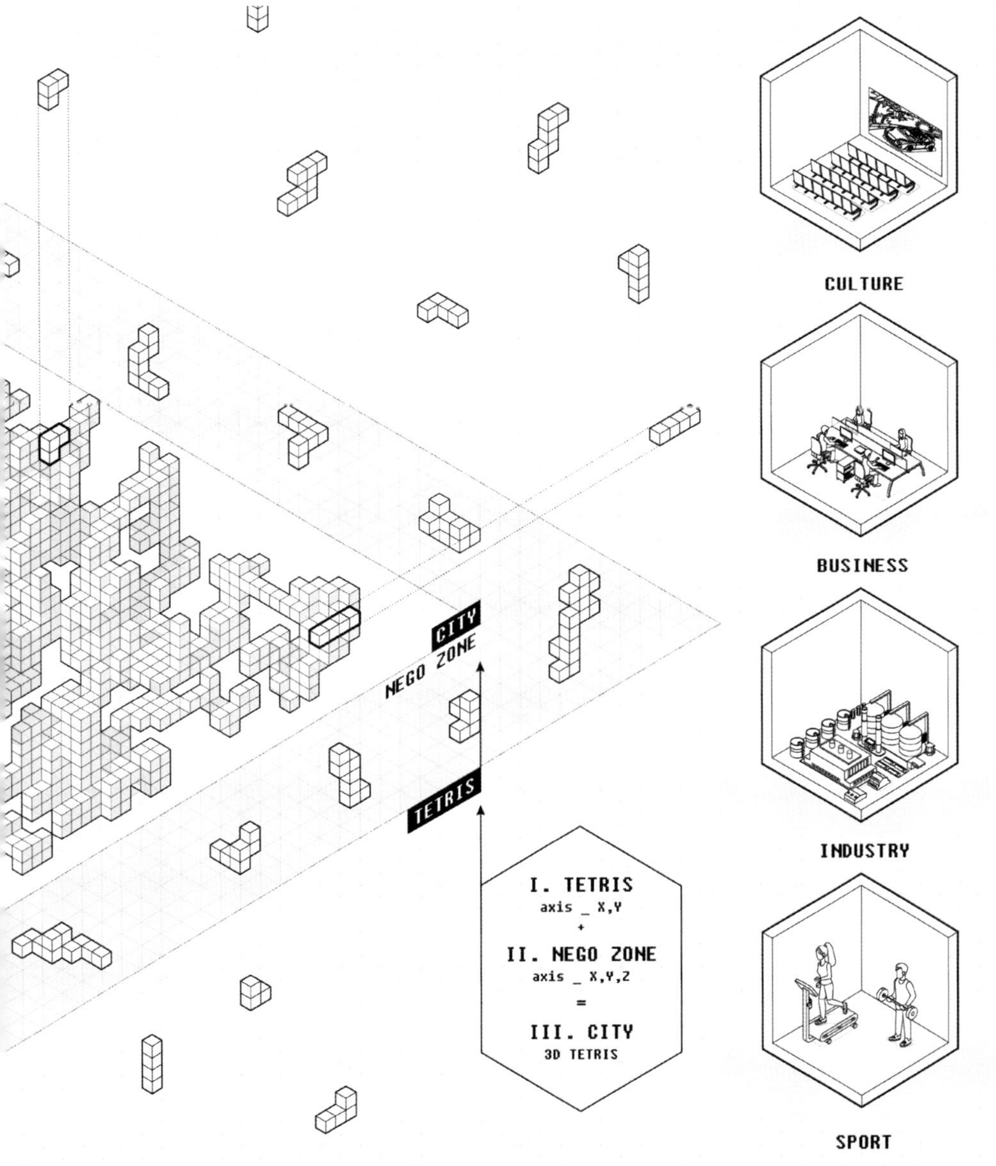

Rulebook of Negotiation city

I. BASIC × II. 3D Zoning × III. Block Placement ×

1. Grid or site

① 기본 도시는 북측에 산지가 위치한 2km 직경의 원형 site로 설정한다.
② Negotiation city는 높은 밀도와 집적된 이용률을 위해 직경 2km의 주어진 site 범위 내에서 600m x 600m로 범위를 제한한다.
③ 하나의 block은 50m x 50m, 보행을 위한 road는 10m 폭으로 구성된다.
④ 그라운드 레벨의 road는 오로지 보행자를 위한 보행로로만 설정한다.
⑤ 600m x 600m 범위로 설정된 기본 도시 이외의 공간은 앞으로의 도시의 확장가능성을 염두에 두고 그리드만 남겨둔다.

2. Zoning

① Negotiation city는 2D 조닝이 아닌 3D 조닝을 기본 원칙으로 한다.
② Zoning box는 10m x 10m를 기본 규격으로 한다.
③ Zoning box는 주거, 상업, 문화, 업무, 교육, 산업, 공공, 농업, 녹지 총 9가지로 구성된다.
④ Zoning box는 기본 모듈(5m x 5m) 8개로 이루어진다.

3. Module

① Module은 기본적으로 5m x 5m를 기본 규격으로 frame과 panel로 구성된다.
② Module을 구성하는 panel은 module이 이루는 프로그램의 성격에 따라 유연하게 변경하여 사용할 수 있다.
③ Module을 지탱하는 frame과 panel은 재활용이 가능한 친환경 소재를 사용한다.

4. Mobility

① 40m level과 70m level에 mobility의 이동을 위한 mobility road과 위치한다.
② Ground level에 위치한 grid를 기본으로 40m, 70m level에서 각각 110m, 230m 간격의 grid를 형성한다.
③ Negotiation city 내 mobility는 개인의 소유가 아닌 공유 mobility를 원칙으로 한다.

I. BASIC × II. 3D Zoning × III. Block Placement ×

1. Proportion of Program

	Residential	Commercial	Cultural	Business	School	Industrial	Institution	Farming	Open space
Proportion	26.9%	10%	6.9%	6.9%	4.6%	6.9%	6.9%	3.8%	26.9%

2. Distance of program

	Residential	Commercial	Cultural	Business	School	Industrial	Institution	Farming	Open space
Residential	O	O	O	O	O	X	O	O	O
Commercial	O	O	O	O	O	O	O	O	O
Cultural	O	O	O	X	O	X	X	O	O
Business	O	O	X	O	X	O	O	O	O
School	O	O	O	X	O	X	X	O	O
Industrial	X	O	X	O	X	O	X	O	O
Institution	O	O	X	O	X	X	O	O	O
Farming	O	O	O	O	O	O	O	O	O
Open space	O	O	O	O	O	O	O	O	O

I . BASIC × II . 3D Zoning × III. Block Placement ×

3. Typology of combination

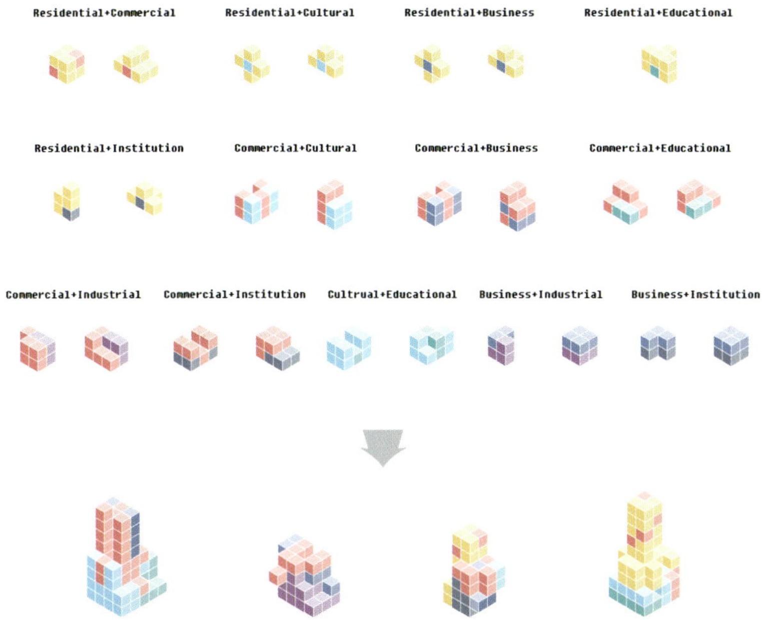

I . BASIC × II . 3D Zoning × III. Block Placement ×

1. Rule for placement

① 학교는 접근성을 고려해 최대한 Ground level에 배치한다.
② Institution은 mobility road의 core 역할을 위해 인근에 우선 배치한다.
③ 일조권과 보행권을 고려해 50m 간격에 필수로 Open space를 배치한다.
④ Commercial은 목적성을 가진 선형적 배치를 통해 상업적 성장을 장려한다.
⑤ Educational은 Natural green way와 인접하게 우선 배치한다.
⑥ Cultural, Institution, residential은 최대한 골고루 분포될 수 있게 배치한다.

2. Rule for combination

① 결합되었을 시 효율이 가장 높은 이상적인 형태의 결합을 원칙으로 한다.
② 회전 시 동일한 형태일 경우 한가지 결합으로 본다.
③ Residential은 필수로 최소 두면 이상이 open space와 접해야 한다.
④ Residential, Educational은 Industrial과 최소 50m 이상의 간격을 두고 결합한다.
⑤ Cultural, Business, Educational, Industrial은 같은 군 zoning box 최소 2개 이상이 결합됨을 원칙으로 한다.
⑥ 학교는 최소 12개 이상의 zoning box가 결합되어야 한다.
⑦ 학교는 운동장 용도로 사용되는 12개 이상의 open space zoning box와 인접하는 것을 우선시한다.

Scenario

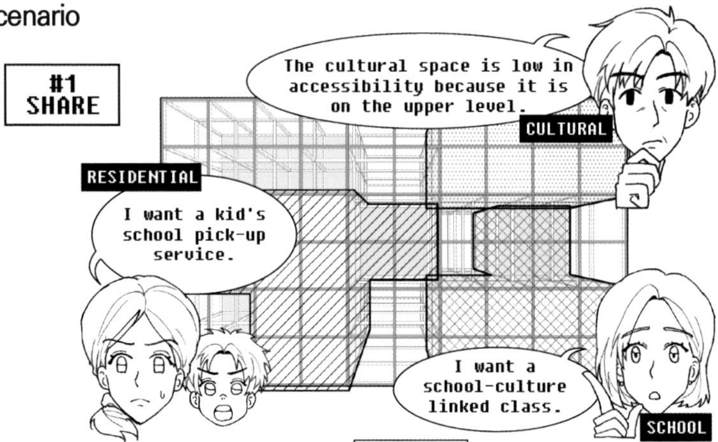

NEGO START!

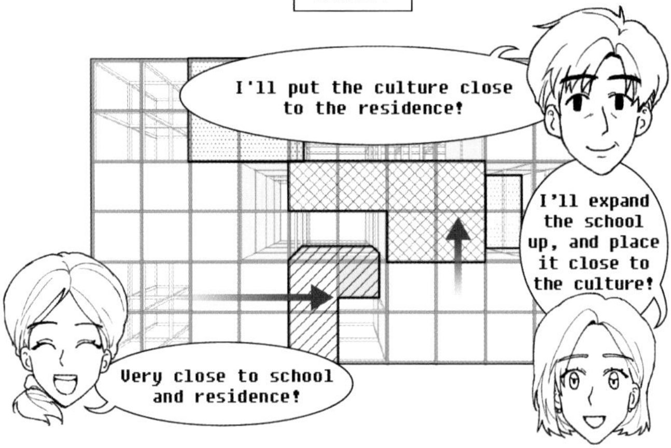

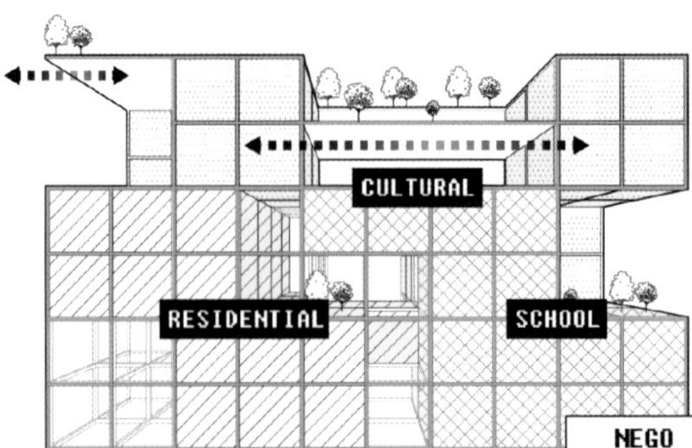

* Connect to green walkway

NEGO FINISH!

#2 ADD

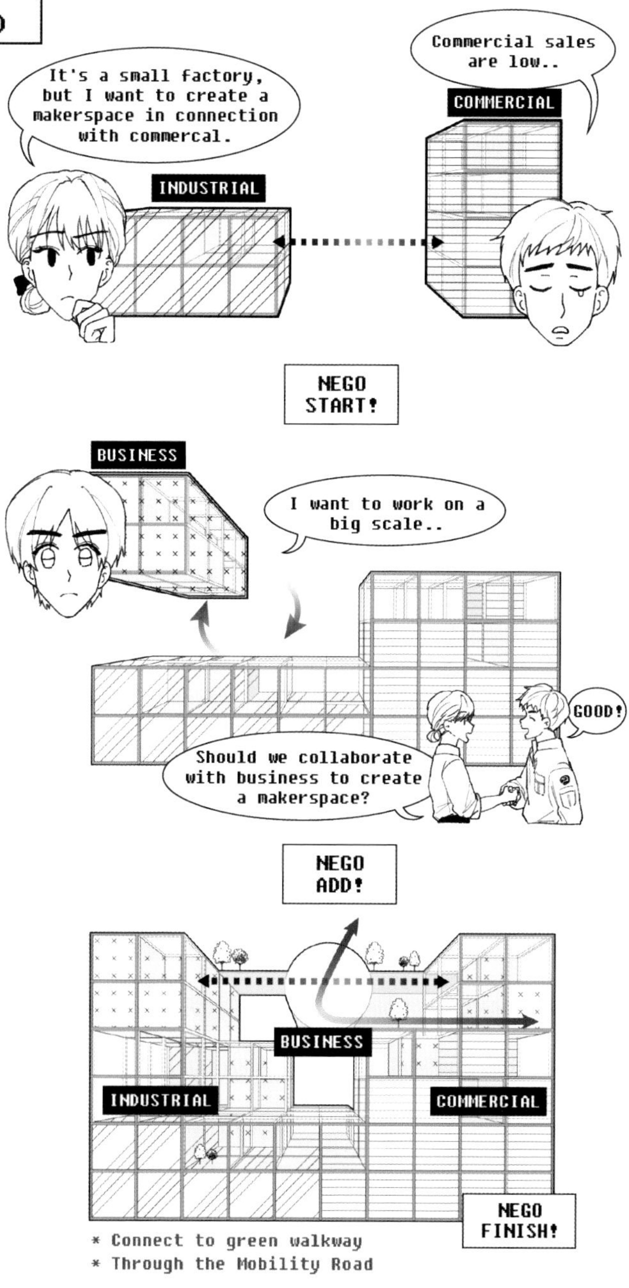

It's a small factory, but I want to create a makerspace in connection with commercal.
Commercial sales are low..
INDUSTRIAL
COMMERCIAL
NEGO START!
BUSINESS
I want to work on a big scale..
GOOD!
Should we collaborate with business to create a makerspace?
NEGO ADD!
BUSINESS
INDUSTRIAL
COMMERCIAL
NEGO FINISH!
* Connect to green walkway
* Through the Mobility Road

Modular system

1. Module & Frame system

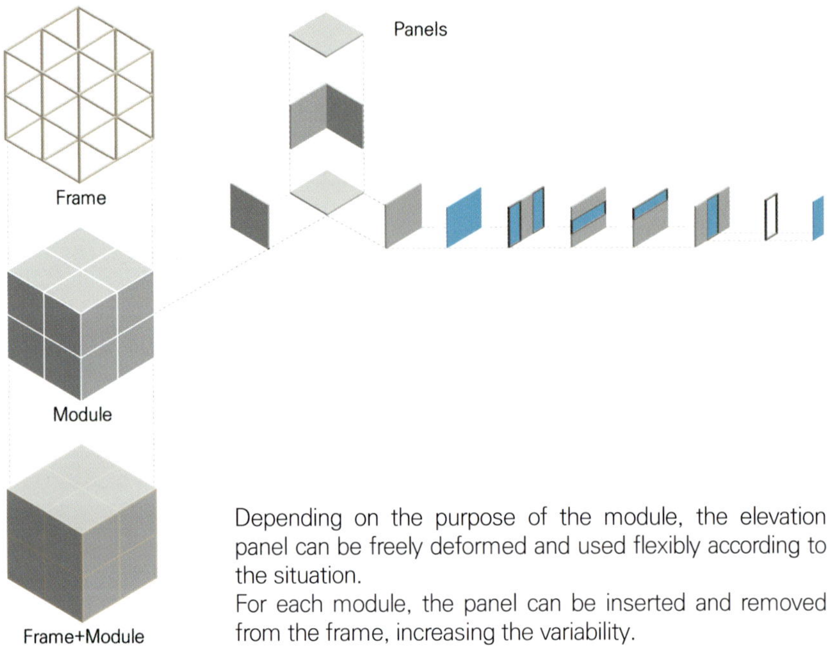

Depending on the purpose of the module, the elevation panel can be freely deformed and used flexibly according to the situation.
For each module, the panel can be inserted and removed from the frame, increasing the variability.

2. Elevation panels by program

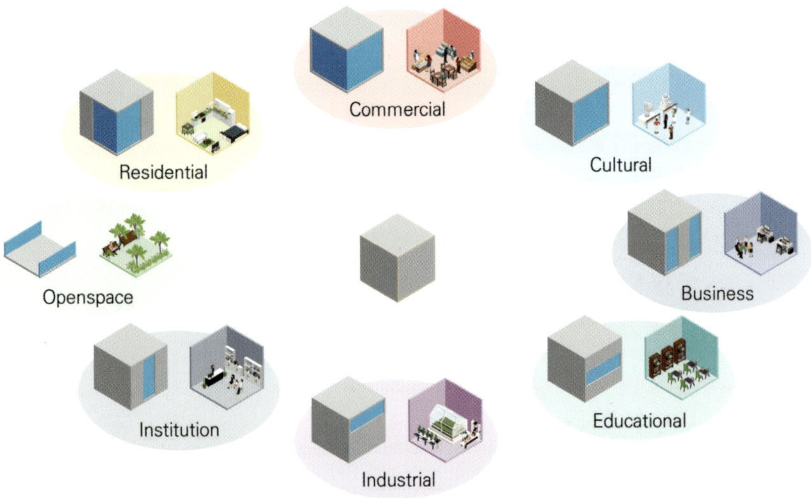

Sunlight Test

Daylighting environments for mid-rise and low-rise residential units

Step 1

Based on the Rule Book, each block is designed to fit a 50m x 50m grid.

Step 2

In order to ensure adequate sunlight for residential units in the mid-rise and low-rise sections, a sunlight simulation was conducted from 10 AM to 4 PM daily, from the winter solstice to the summer solstice. This allowed us to identify the units that reduce sunlight exposure to the residential units.

Step 3

By removing or relocating the specific units, we were able to ensure that the sunlight conditions for all residential units within the block met a certain standard. After simulating from all four directions at the block level, the final massing was determined.

Program Distribution

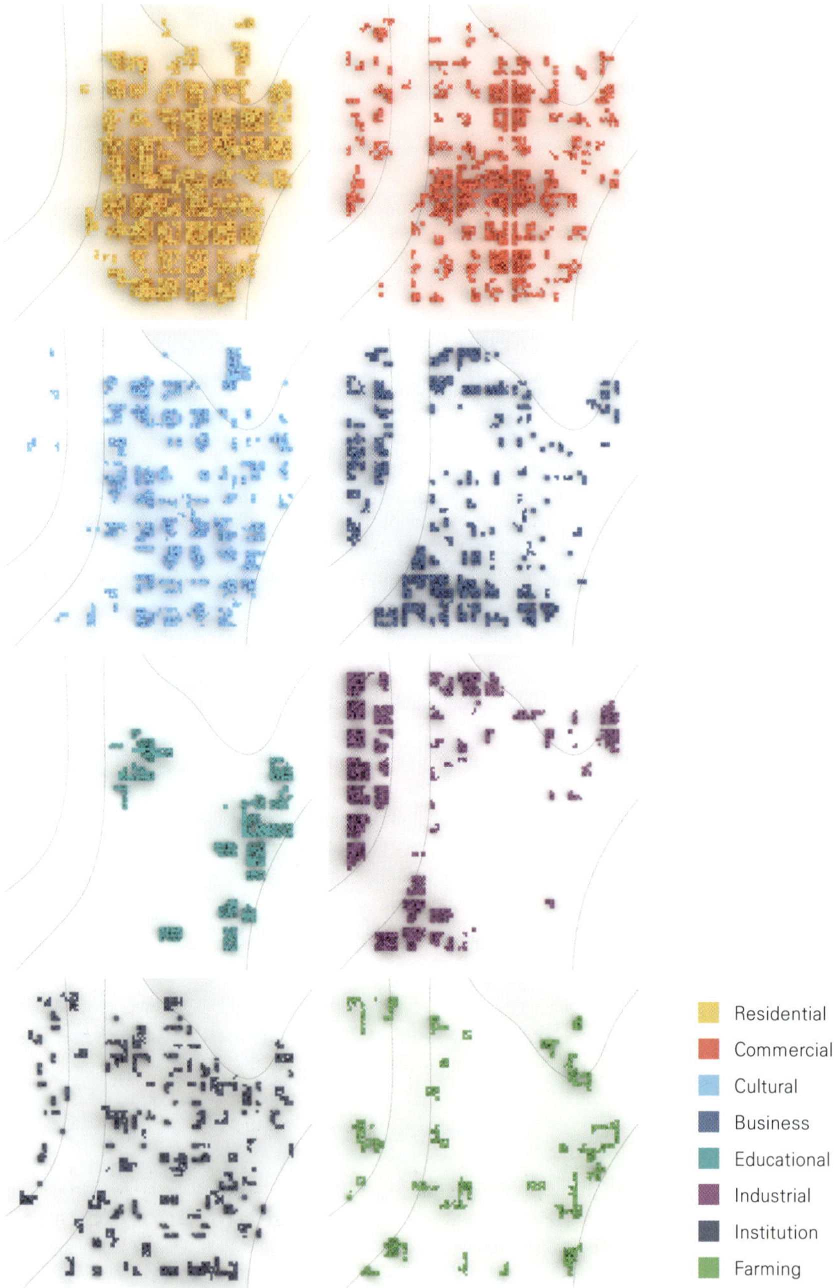

Program Connectivity

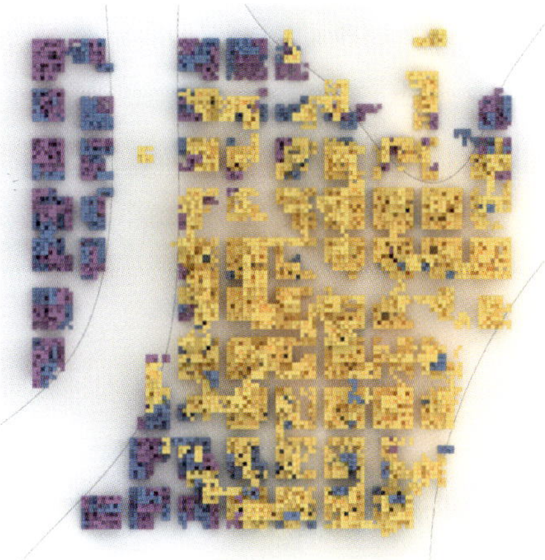

Residential + Business + Industrial

Residential areas are distributed separately from industrial programs such as factories, whereas office spaces are often linked to industrial areas based on their respective concerns.

Educational + Farming

Educational facilities are arranged to be connected with as many agricultural facilities (such as smart farms and green spaces) as possible, aiming to encourage activities linked with open spaces.

Commercial + Cultural

In the case of commerce, commercial streets form a cross shape at the city center, while small-scale commercial facilities are distributed throughout the city. Cultural spaces are usually integrated around these areas to ensure connectivity with commercial activities.

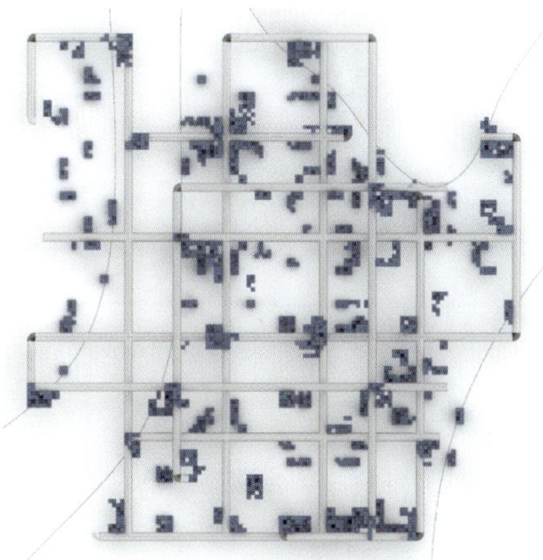

Institution + Mobility road

Various public facilities are located around roads, serving as stops for mobility. Additionally, small public modules are dispersed throughout the city to provide convenience in daily life.

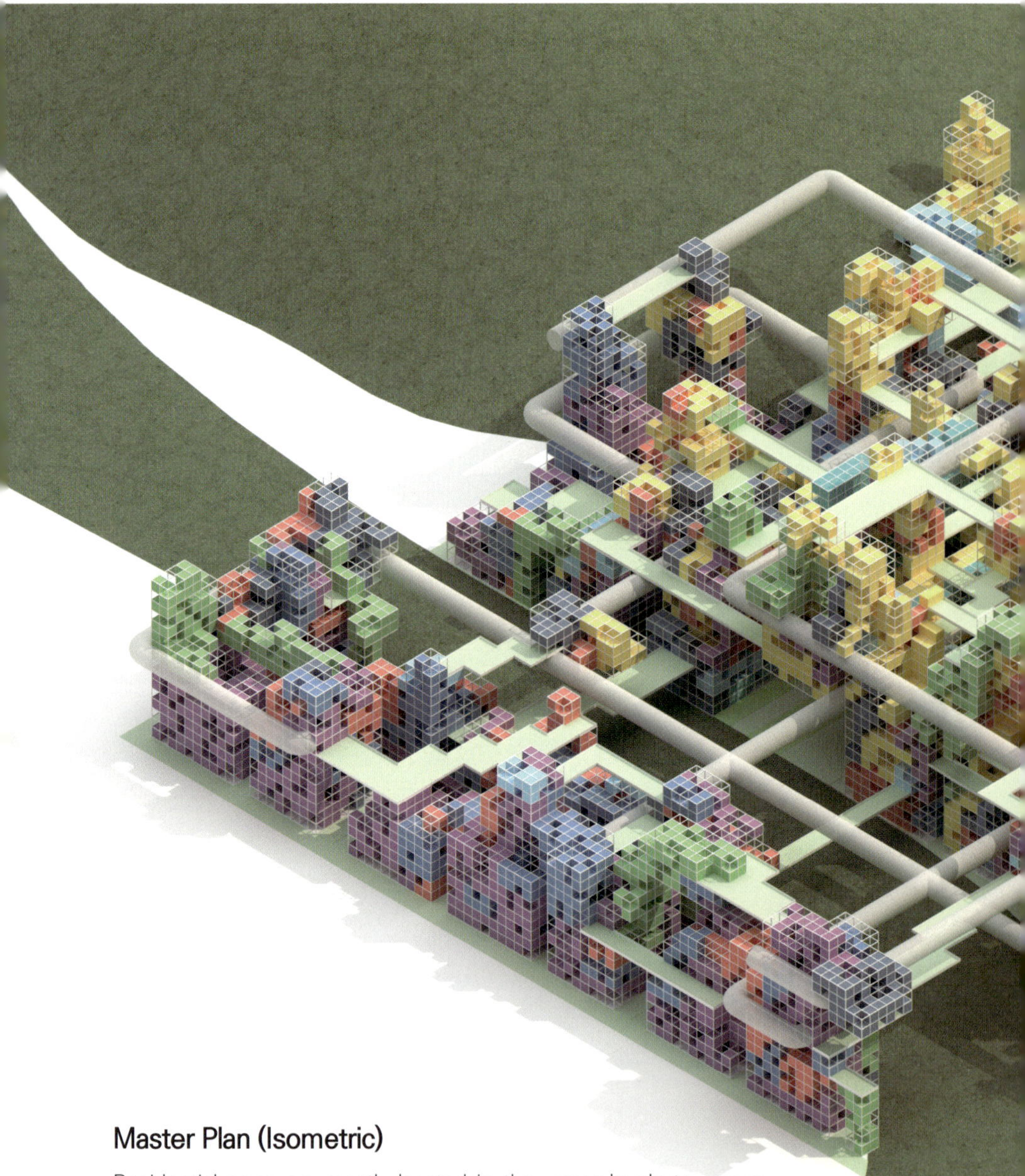

Master Plan (Isometric)

Residential areas are mostly located in the upper levels to ensure adequate sunlight, and they are integrated with commercial, cultural, and public facilities. Additionally, elevated walkways connecting the various modules will allow residents to move efficiently through an ideal environment, not only relying on mobility but also through walking.

Master Plan (Isometric)

This is the envisioned appearance when the Negotiation city is actually formed. Through the negotiation between residents and nature along the walkways and among the modules, many green spaces will be created in various forms throughout the city. The city will be fulfilled with individual needs while simultaneously creating an ideal environment.

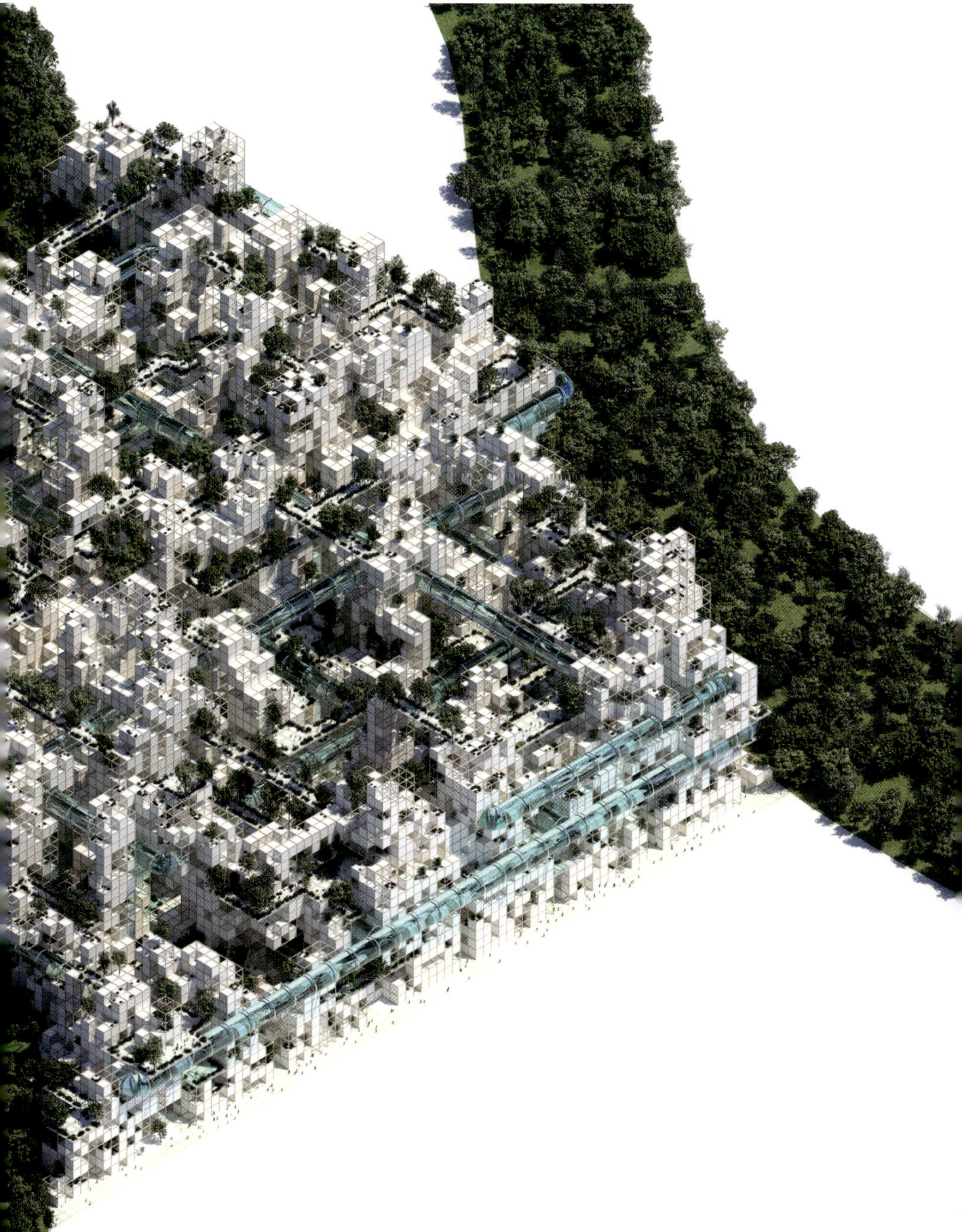

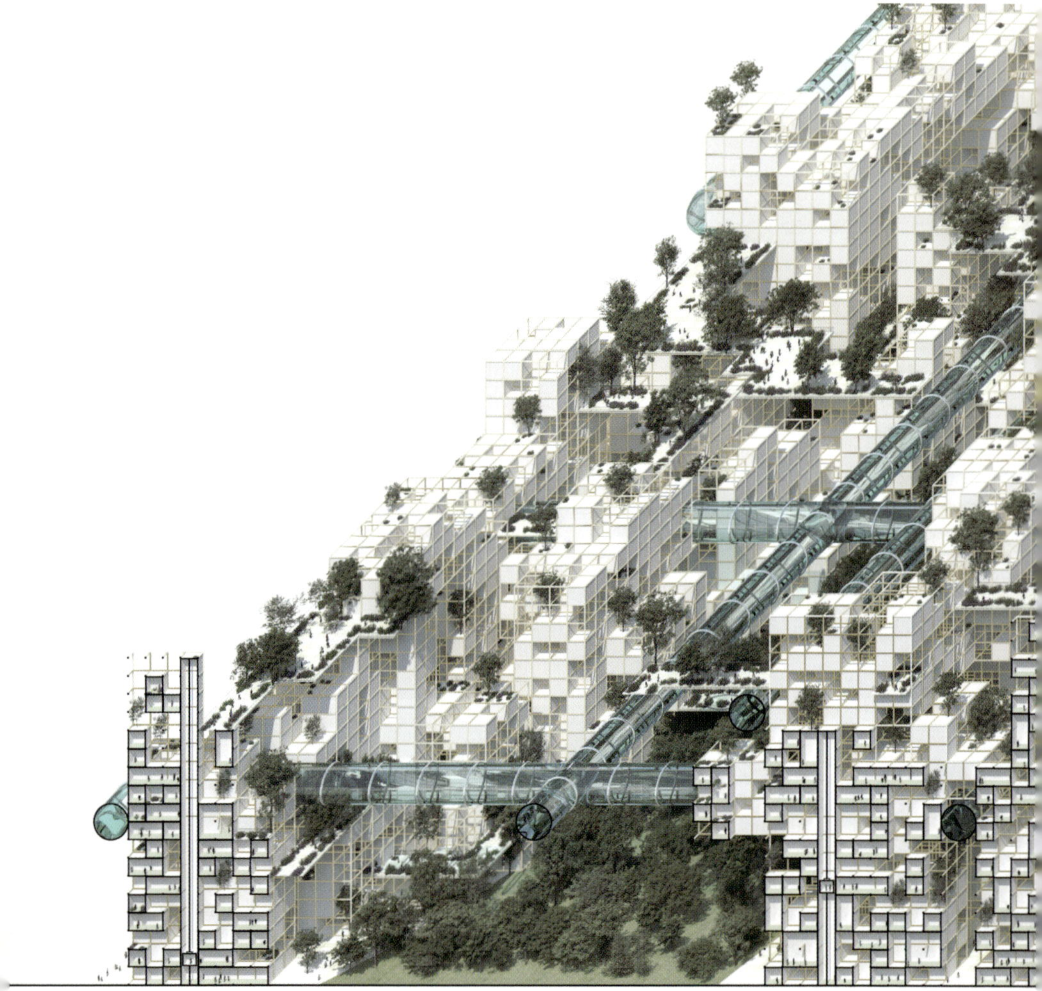

Oblique View Section View

Each module does not end with the basic layout of 5m x 5m; it can be expanded horizontally and vertically based on the characteristics of the program and the needs of the users. Vertical movement within the modules can be achieved using elevators installed within the block, allowing access to the desired floors. For movement between different blocks, elevated walkways or ground-level pathways can be used. In this way, Negotiation City can facilitate movement and spatial configuration according to the needs of each user.

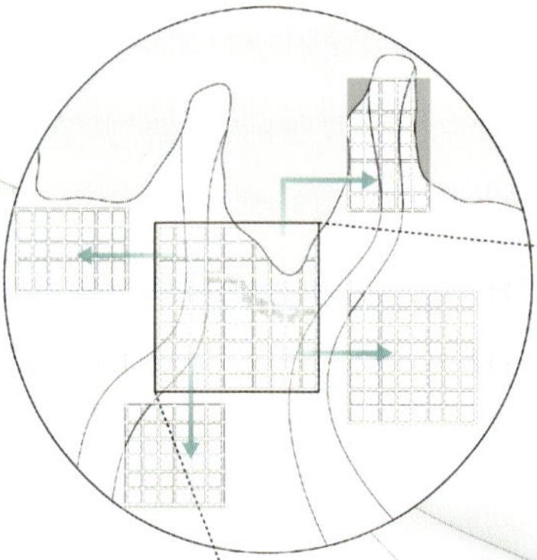

The city is endlessly expandable along the existing grid(50m50m).

The city is expanded, can be negoed according to the user's lifestyle, and is formed in various forms.

It can also be connected to the existing site grid and mobility load.

1 [EXPANDABILITY MAP]

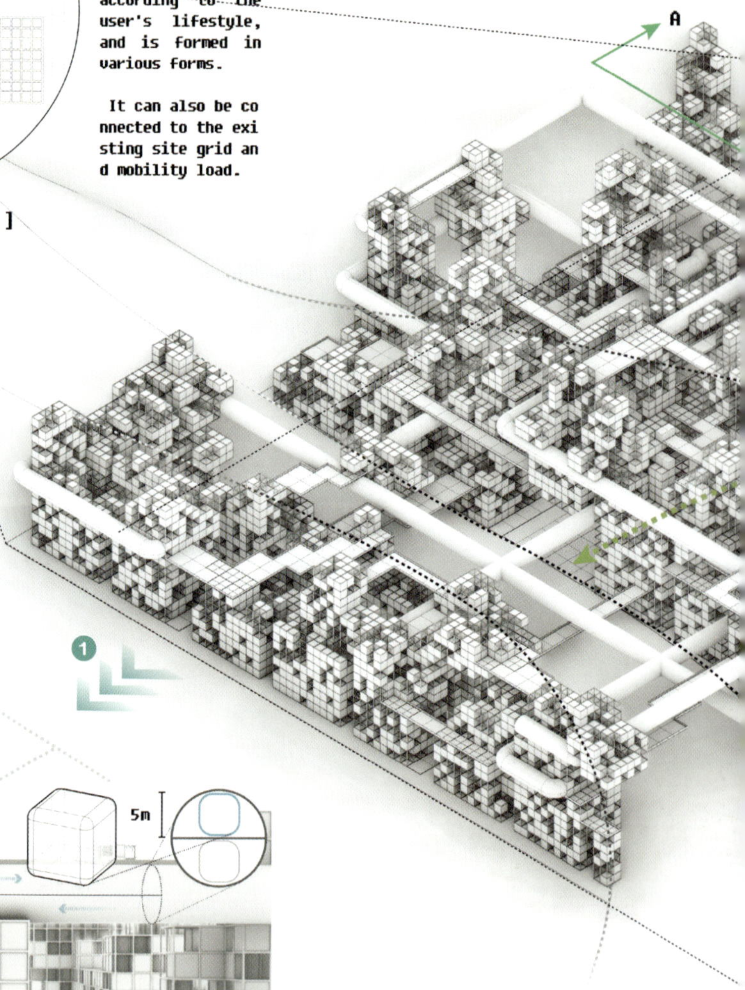

3 [MOBILITY]

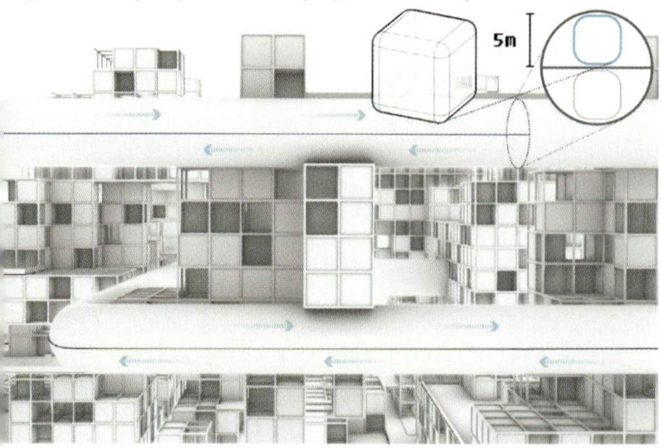

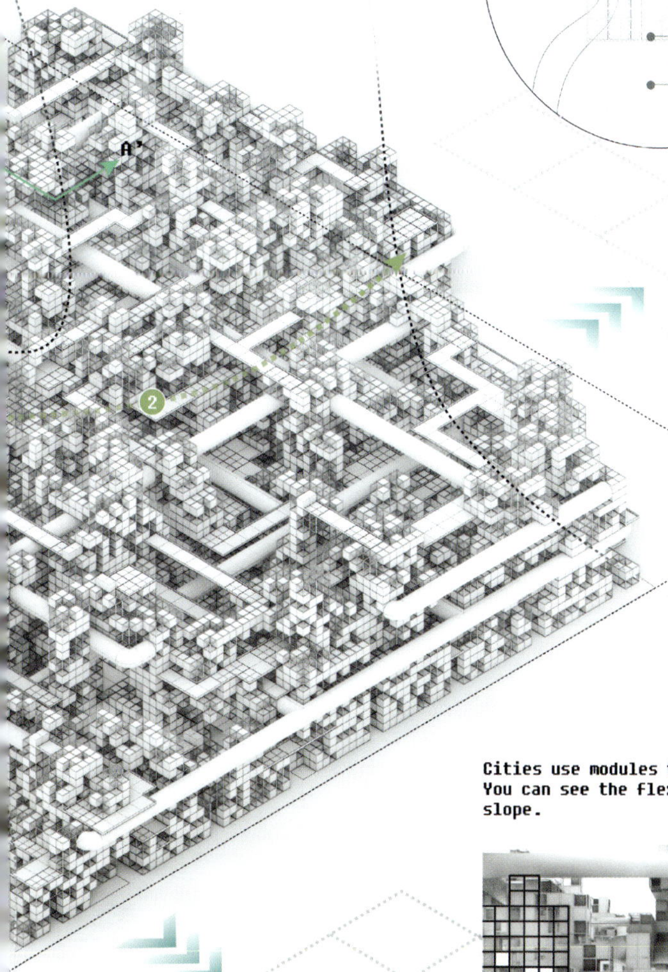

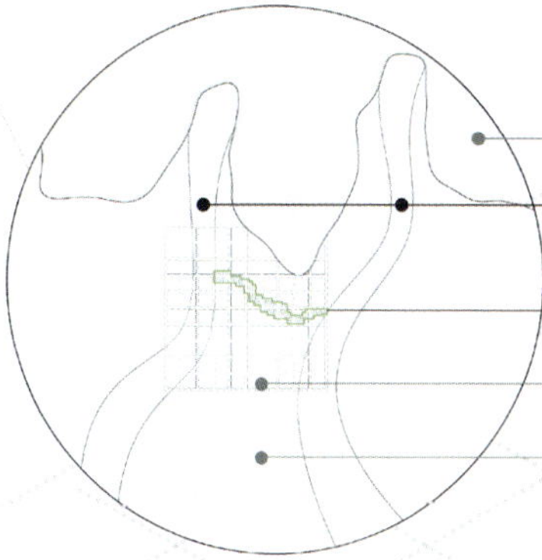

[GREEN ZONE MAP] 2

[A-A’ SECTION] 4

Cities use modules that can be flexibly applied anywhere. You can see the flexible application of the module to the slope.

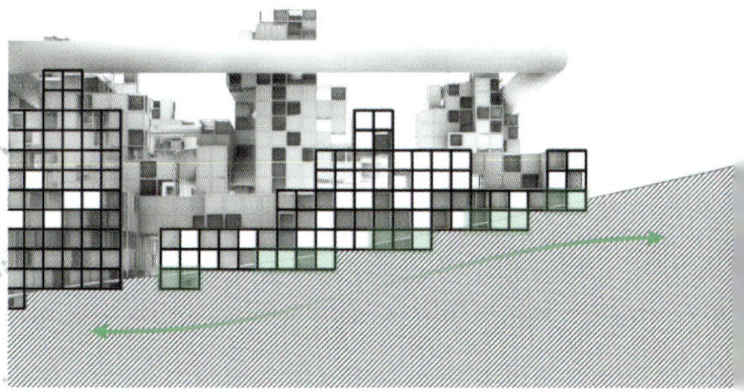

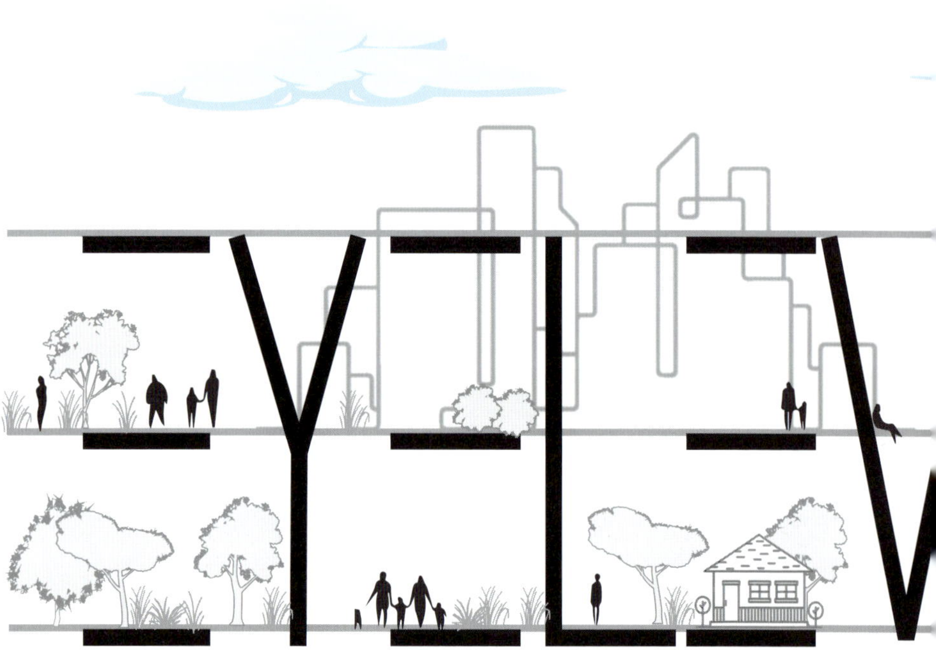

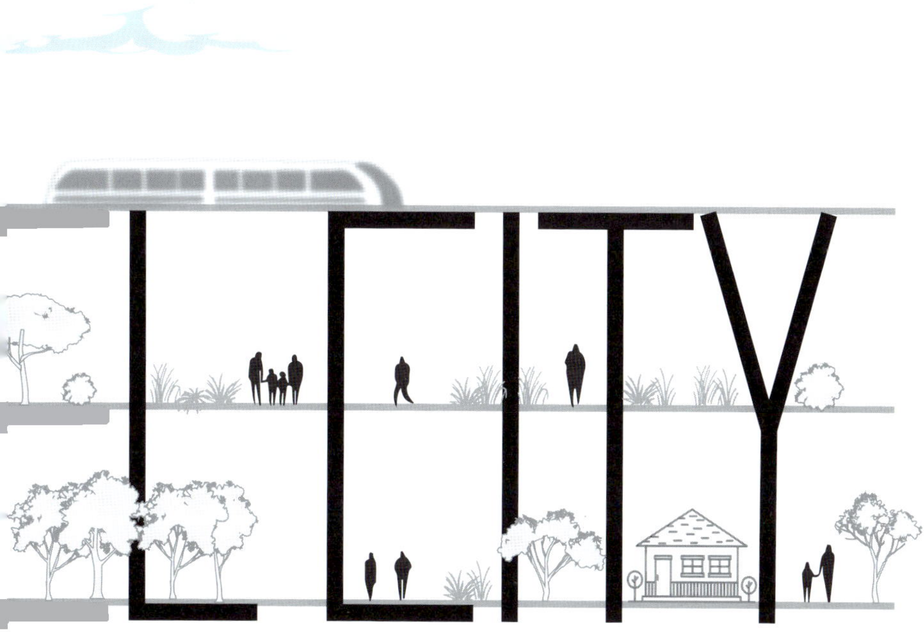
CITY

EYELEVEL CITY

4조 : 문영은, 박상우, Charlaine

The development of cities and the history of humanity are closely intertwined. From the ancient walled city of Rome, to the fortified towns of the Joseon era, and up to the modern metropolis, cities have served as the hub of social, economic, and cultural activities, greatly influencing human life.

The Industrial Revolution further highlighted this development, impacting the expansion of cities. However, such urban development has given rise to various issues such as traffic congestion, environmental pollution, and social disparity. Now, there is a growing call for a return to people-centered, pedestrian-friendly urban development to address these problems.

Examples such as Denmark's Copenhagen, which has expanded bicycle roads and created pedestrian-centric streets to lead to 50% bicycle commuting, Brazil's Curitiba's BRT system that reduces vehicle use and promotes sustainable urban development, and the U.S.'s Portland, which has set city boundaries to limit urban expansion and increase green spaces, demonstrate how cities can transition to people-centered development to solve problems and improve quality of life.

We must learn from these examples and reconstruct our cities for a better future. Cities should not be simply a collection of buildings and roads, but spaces that reflect the quality of life of the people living within them. Therefore, human-centered, pedestrian-friendly urban development should not be a simple choice, but an essential direction.

Background

Existing Urban Plan

The appearance of a city is bound to reflect the technological, social, and cultural characteristics of the time.

It also grows with human history, with ancient cities developing fortresses around temples, and Rome developed infrastructure and public buildings such as roads and waterways.

Medieval cities are represented by churches, monasteries, defensive bastions, and walls. In the case of the East, the city was formed with the traditional residential areas of fortresses.

Such urban development has been structured around the patterns and needs of people's lives, and city development prior to industrialization was human - centered.

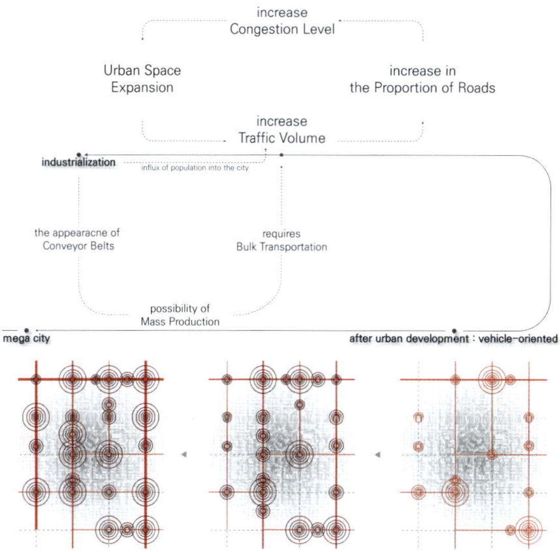

While industrialization has had positive impacts on human life, from the perspective of urban development, this is not the case. Industrialization led to the emergence of the conveyor belt, which led to mass production and created a need for mass transportation. This implied the need for increased road capacity, and as new roads were built, cities expanded, continually drawing people into the city.

This trend in urban development has led to the formation of megacities, which in turn has resulted in problems such as traffic congestion, environmental pollution, and population density.

Characteristic

Strong Small City

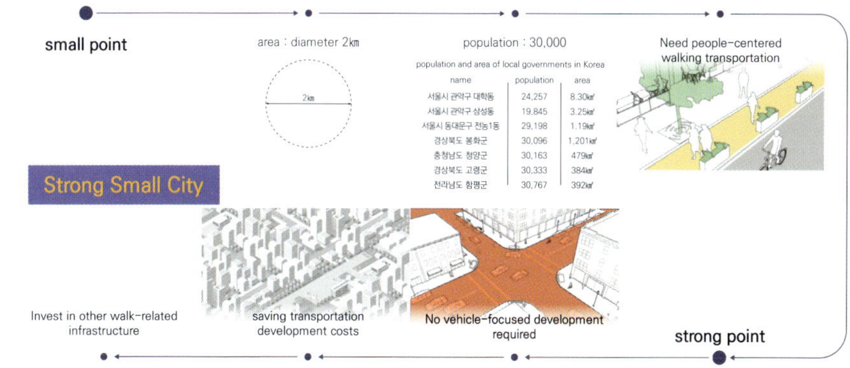

population and area of local governments in Korea

name	population	area
서울시 관악구 대학동	24,257	8.30㎢
서울시 관악구 삼성동	19,845	3.25㎢
서울시 동대문구 전농1동	29,198	1.19㎢
경상북도 봉화군	30,096	1,201㎢
충청남도 청양군	30,163	479㎢
경상북도 고령군	30,333	384㎢
전라남도 함평군	30,767	392㎢

Small cities should utilize their characteristics in this situation. In line with Korea's decreasing population, small cities with smaller populations and sizes need pedestrian-centered transportation. Furthermore, by investing in necessary urban facilities such as pedestrian infrastructure with the cost needed for vehicle-centered development, they can become competitive with megacities.

3rd Pedestrian way
strong small city is
1st Pedestrian way

EYELEVEL CITY

Detail of Manifesto

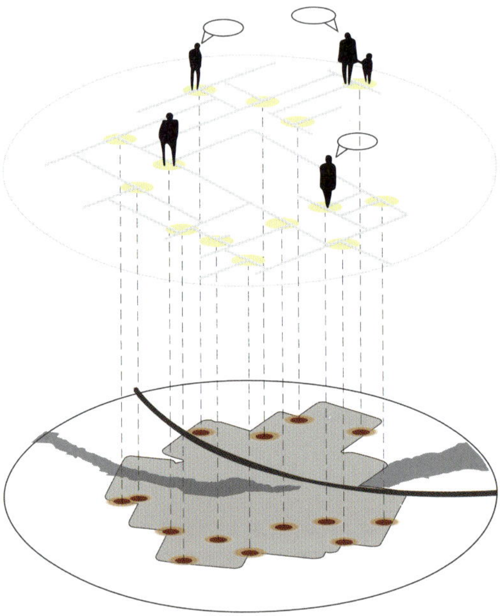

Eye-Level with Communication

This diagram shows a city where various circles represent communication and connection hubs.

These hubs are strategically placed throughout the city, ensuring that interaction between people occurs naturally at the eye level.

This emphasizes the importance of face-to-face interactions and community engagement, fostering a sense of belonging and social cohesion among residents.

Eye-level with Scenery

Representing the integration of eye-level design with scenic elements within the city. This visual showcases a cityscape where the urban environment is harmoniously blended with natural scenery.

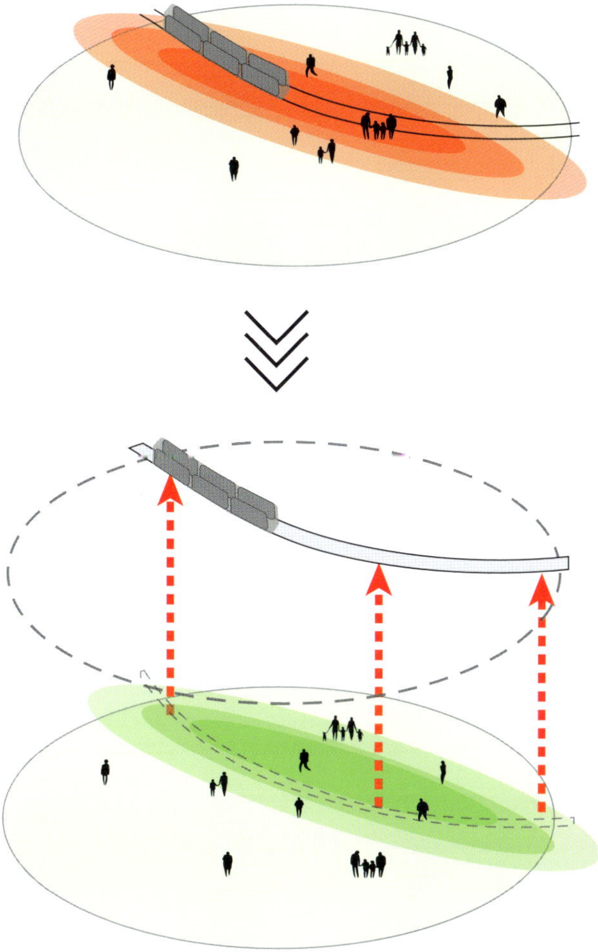

Eye-Level under Mobility

By elevating transportation systems such as monorails above the ground, we preserve ground-level views and ensure that streetscapes remain open and inviting. This design choice allows various urban uses, including housing, commerce, and green spaces, to be easily accessed within the pedestrian's visual range.

Furthermore, elevating the monorail significantly enhances pedestrian safety by reducing the risk of accidents with ground-level traffic. The result is a more cohesive and connected urban environment where transportation infrastructure supports rather than disrupts the urban fabric.

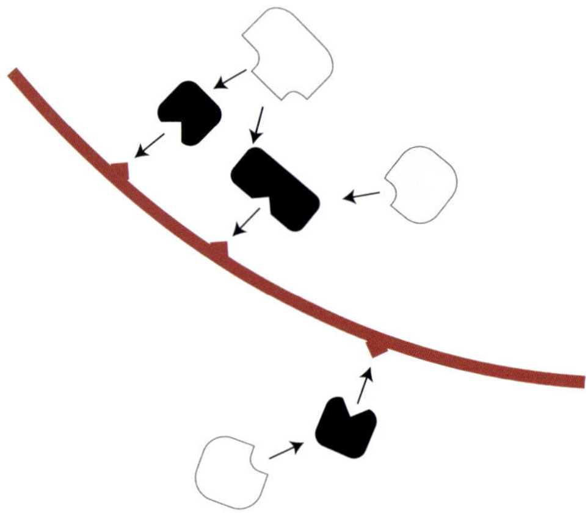

The Plug-in System

The Plug-in System enables flexible, small-scale developments for smaller compact urban areas. In the initial phase, plug-ins which are the communities are strategically connected to the elevated monorail, ensuring seamless integration with the city's transportation infrastructure. The other communities will plug into the first plug. The communities road boundaries, when plugged together forms T-junction roads, and several plug-ins collectively form a mat layer. Additionally, the system conserves nature by strategically placing smaller units around existing natural features, minimizing environmental disruption compared to traditional large-scale developments.

Large scale development

Plug-in units stregically placed

Benefits of T-junctions

Safety

In a typical crossroad, numerous intersecting paths increase the likelihood of pedestrian collisions, especially during peak hours.
T-junctions simplify traffic flow by reducing intersecting paths, minimizing conflict points, and significantly enhancing pedestrian safety.

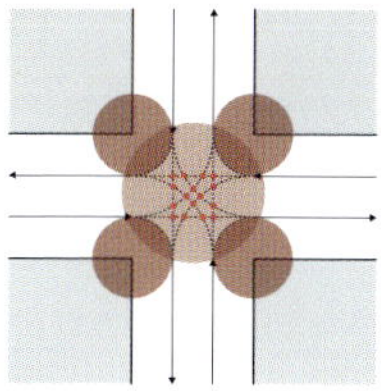

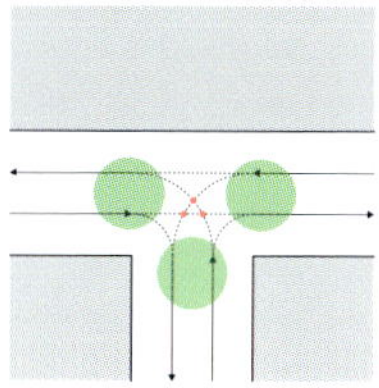

Road area Ratio

T-junctions improve the road area ratio by using space more efficiently than crossroads. This efficient use frees up substantial areas for parks, gardens, and green spaces, enhancing the urban environment's aesthetic appeal and providing valuable recreational spaces for residents.

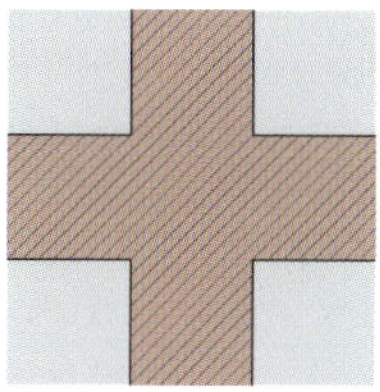

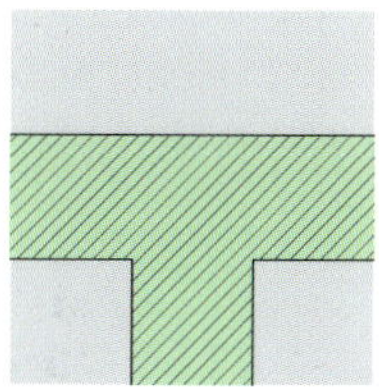

Aesthetic and Navigational Benefits

Crossroad intersections often present monotonous sceneries, lacking visual diversity. In contrast, T-junctions offer varied and aesthetically pleasing designs, with unique landscaping, art installations, and architectural features.
They serve as effective pathfinders, guiding pedestrians intuitively and enhancing the city's charm and navigability.

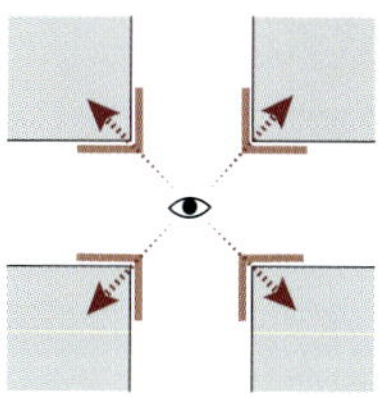

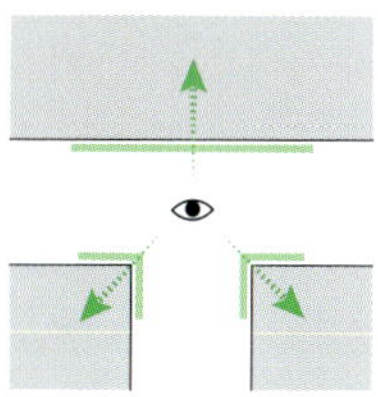

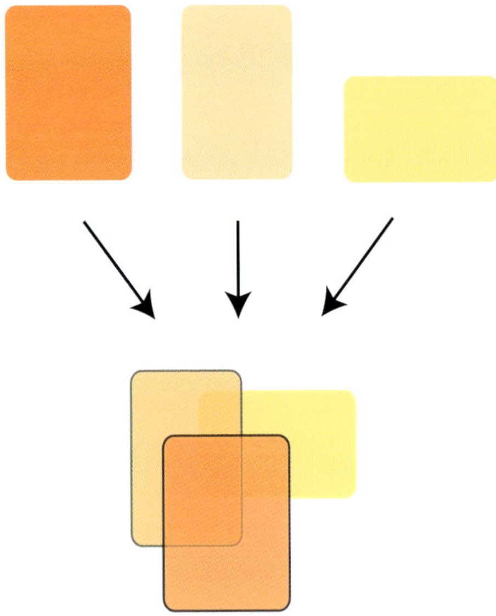

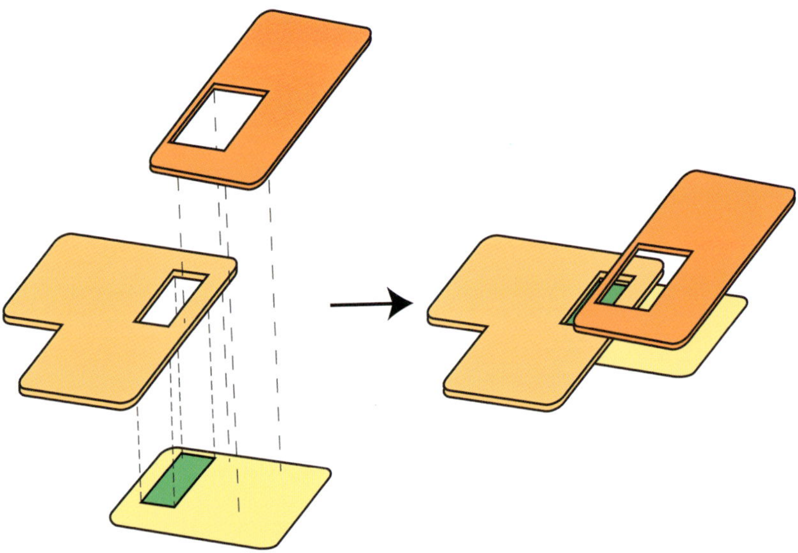

The Mat system

Mat System involves layering multiple mat layers on top of each other. Each mat layer is designed with several voids to ensure ample sunlight and ventilation reach all levels, creating a healthy and pleasant environment for residents and visitors.

The voids within each mat layer are strategically organized to allow views of open green spaces, buildings, and pedestrian paths from the layer above. This design ensures that the urban landscape remains visually connected and accessible, enhancing the sense of community and interaction among different layers. The placement of these voids allows natural light to penetrate deep into the mat structures.

Furthermore, the arrangement of the layers is such that residents and visitors can enjoy panoramic views of the entire mat system when standing by the edges or near the voids. This visual connectivity fosters a feeling of closeness and integration, as individuals can observe and appreciate the diverse urban programs and building typologies from different vantage points. The ability to see and experience different layers from above not only enhances the aesthetic appeal but also provides a unique and engaging urban experience.

Masterplan

We designed a space using Mat system and Plug-in system.
A mat divided into three layers was made, and the plug-in system was reflected with the TOD & Wetland line.

1st Mat

Mat in first stage of EYELEVEL CITY, the formation of basic dwellings and community spaces.

2nd Mat

In addition to the basic mat, educational, general commercial, and cultural functions are added.

3rd Mat

As a final step, work spaces such as shared office spaces and central commercial facilitied are added.

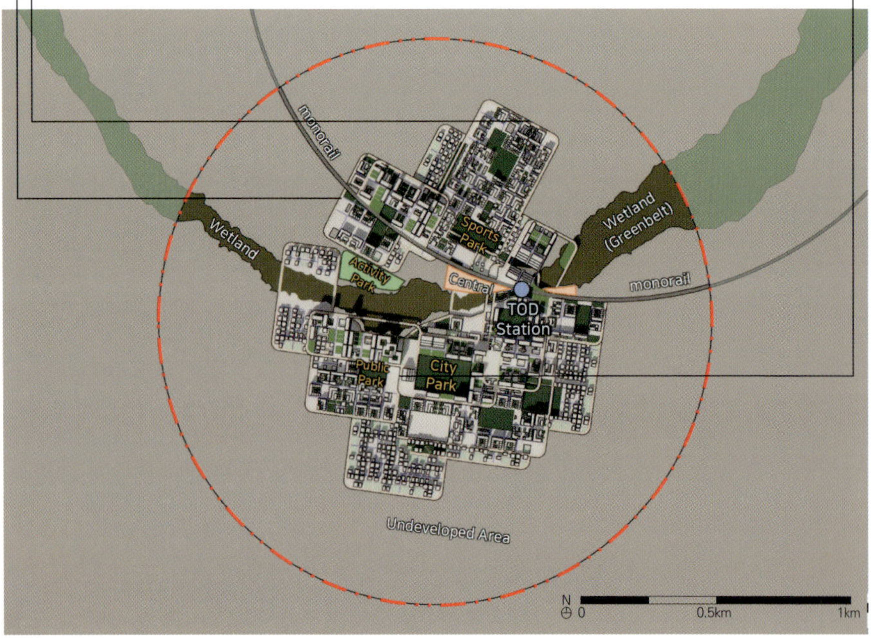

Spatial Plan _ Mat System

1st Mat

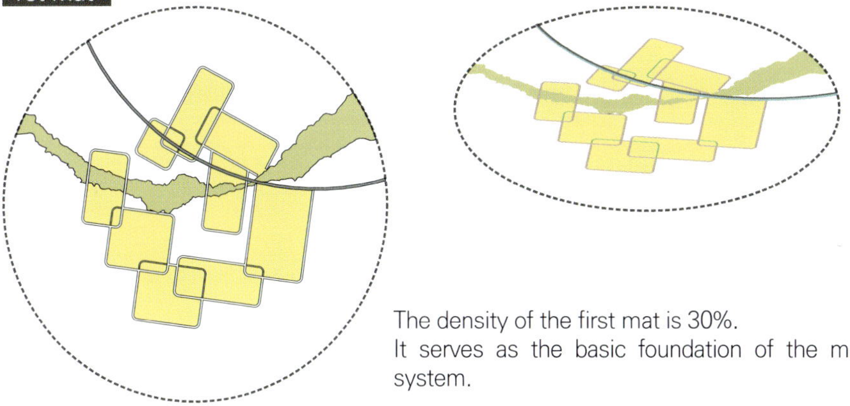

The density of the first mat is 30%.
It serves as the basic foundation of the mat system.

2nd Mat

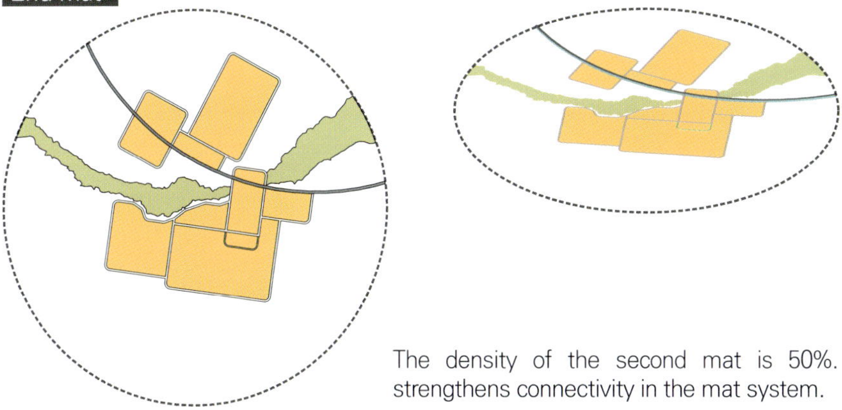

The density of the second mat is 50%. It strengthens connectivity in the mat system.

3rd Mat

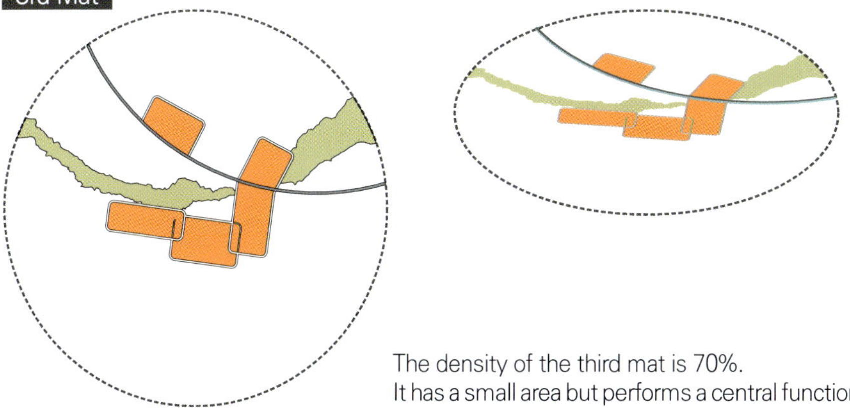

The density of the third mat is 70%.
It has a small area but performs a central function.

Spatial Plan _ Zoning

Zoning Layer

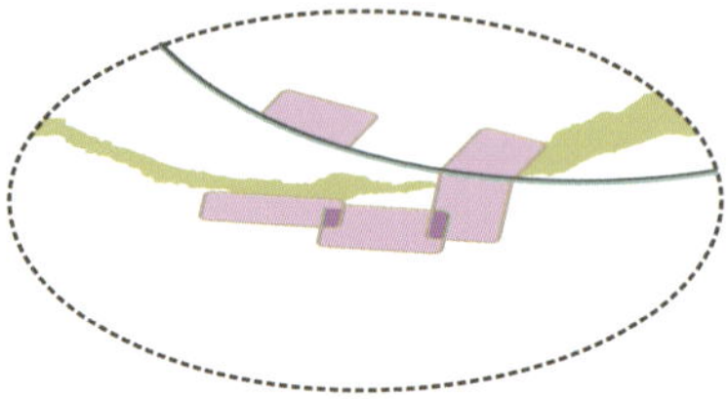

The third mat is a small area but high density, where commercial & institution is located.
It is located near the central TOD for its function and accessibility of the city.

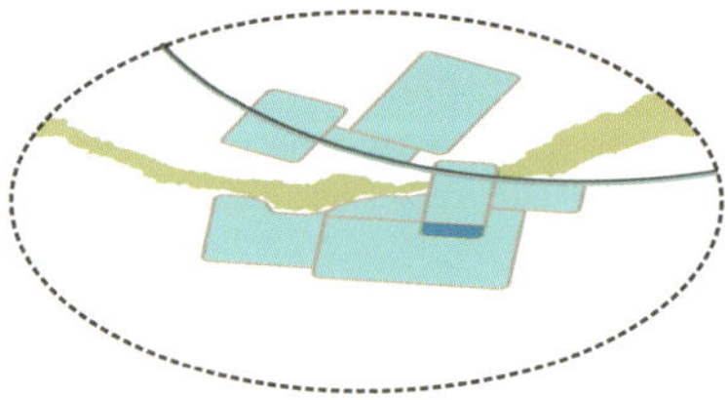

The second mat has a relatively open culture & education building and program. The school can be located and creates an active atmosphere for children's outdoor activities.

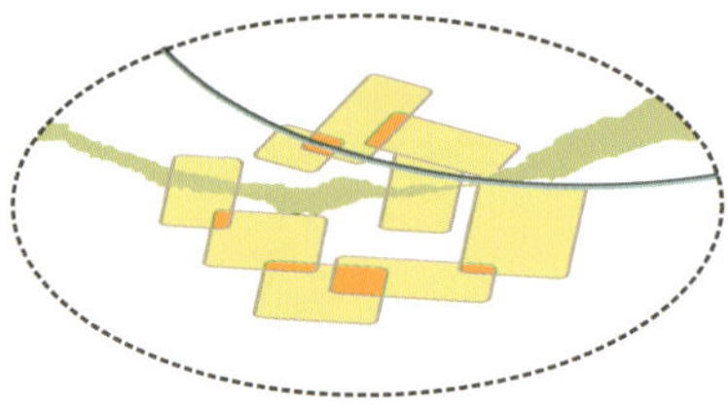

The first mat has residential & community buildings and programs centered on neighborhood.
Due to its large area but low density, people can protect their personal lives and communicate with their neighbors at a proper distance.

Program

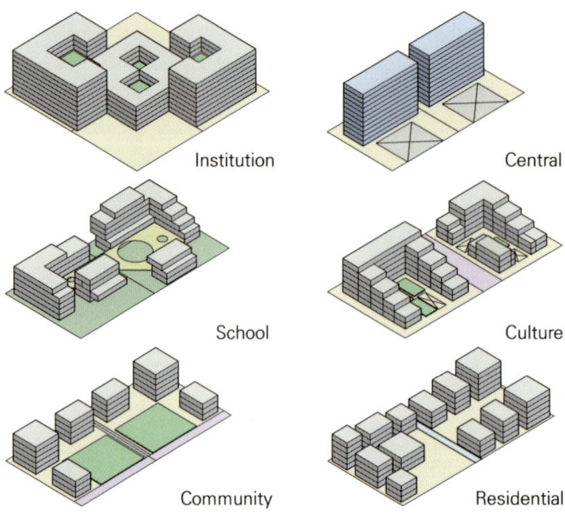

3rd Mat: 11 floor
2nd Mat: 5~8 floor
1st Mat :3~5 floor

Building layout to suit the characteristics of each mat

Diversifying Building Types according to the characteristics

Spatial Plan_Openspace

Wetland

We assumed that wetlands are located adjacent to the TOD line.
When wetlands are located around TOD, they can alleviate environmental pollution that may occur due to complex traffic and act as a countermeasure against unpredictable natural disasters.

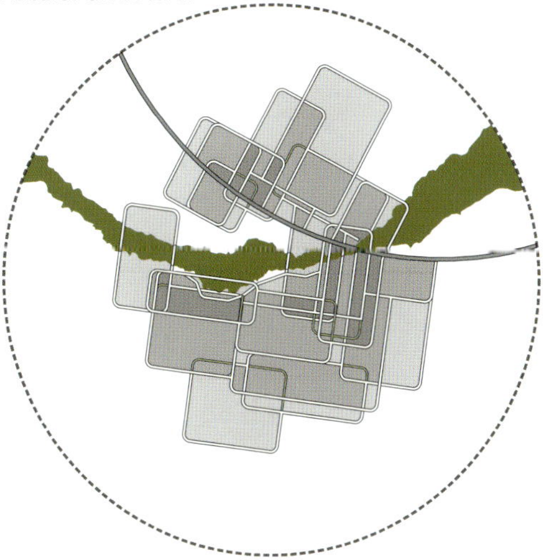

In addition, living things living in wetlands can promote ecological value in small and medium-sized cities and provide people with ecological experience programs.

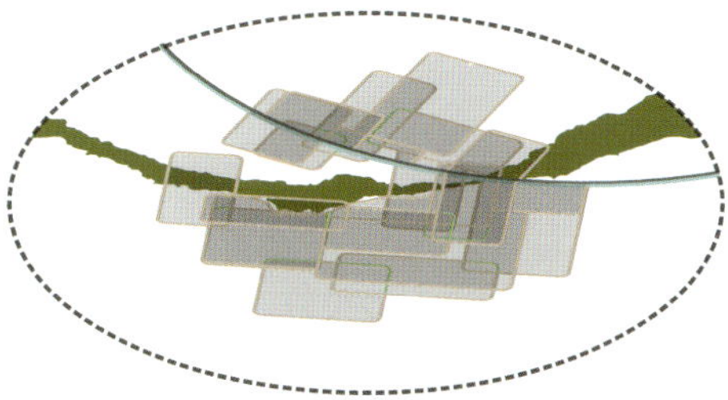

Since frequent human contact can adversely affect the wetland, we did not place the mat system and wetland close. When it was necessary to move across the wetland for efficient movement, direct contact was minimized by installing bridges on the second and third mats.

Wetlands will not only respond to disasters in small and medium-sized cities and purify the polluted environment, but will also play a central role in programs that allow people to experience the natural environment and nature.

Spatial Plan_Openspace

Park

We arranged some kinds of public park in the section where mats on each floor overlap. This park is different from wetlands that minimized contact with people, and is the most open, social, and open space where people come into contact frequently.

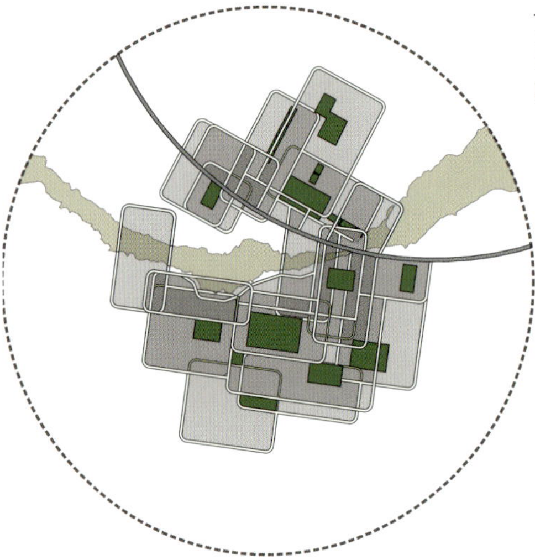

These public parks allow people in eye level cities to harmonize together and promote large communities.

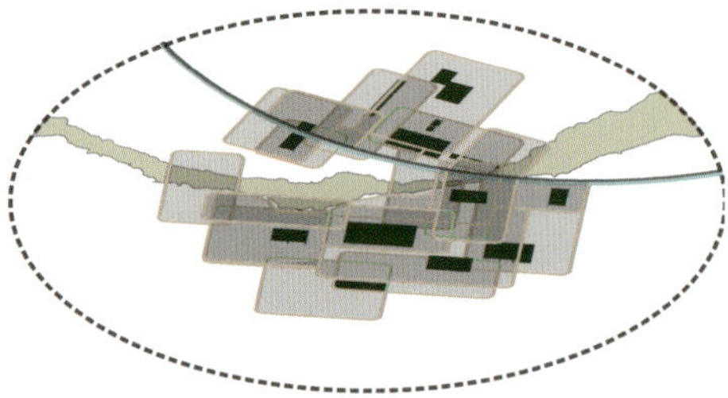

In eye level city, walking is the main means of transportation, so we need a space where we can experience programs through people's walking.So we devised cultural programs such as sports, high-activity, square, and fly market and reflected each program in an appropriate location. The location and size were adjusted differently according to the nature of each program.

The public park provides people in eye level city with a place of meeting and harmony, and is the most open and active space that allows people to experience various programs while walking.

Spatial Plan_Openspace

Pocket Park

We placed a pocket park in the free space of each building. Pocket park is the most private space among open spaces, and is a space that functions as a garden mainly in residential complexes.

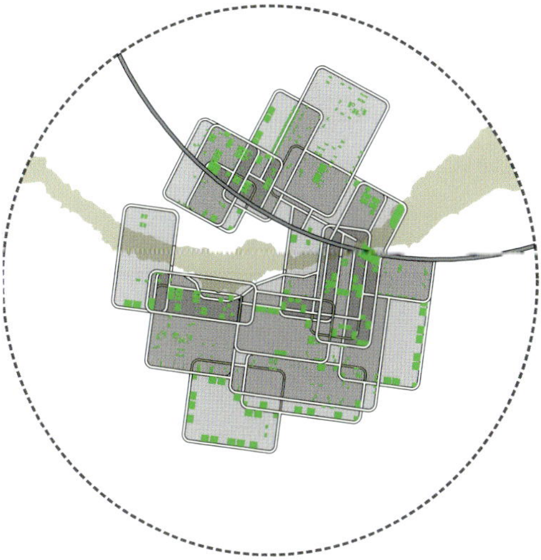

Wetlands have little contact with people and if the park is an open space centered on communities with people, pocket park is a space where contact with neighbors is mainly made.

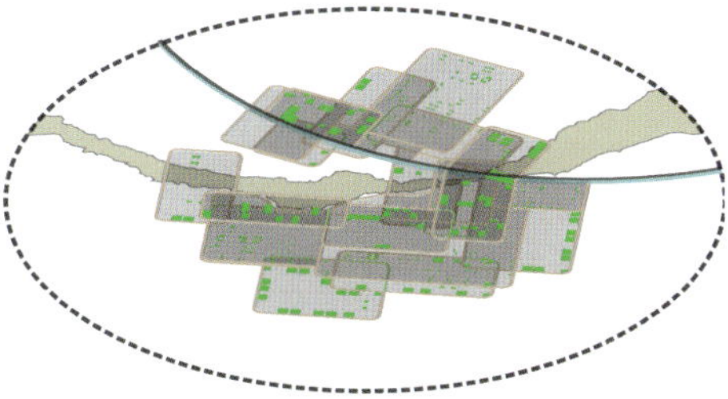

Although the pocket park is a relatively smaller area than other open spaces, it is spread throughout the eye level city.

This can provide people with enough rest in the walking-oriented eye level city, and they can enjoy an outdoor rest near their home and communicate with their neighbors even if they don't go far.

The pocket park is the smallest area but the most distributed, and will serve as a garden for people to relax near their homes and have small communications with their neighbors.

Spatial Plan_Pedestrian

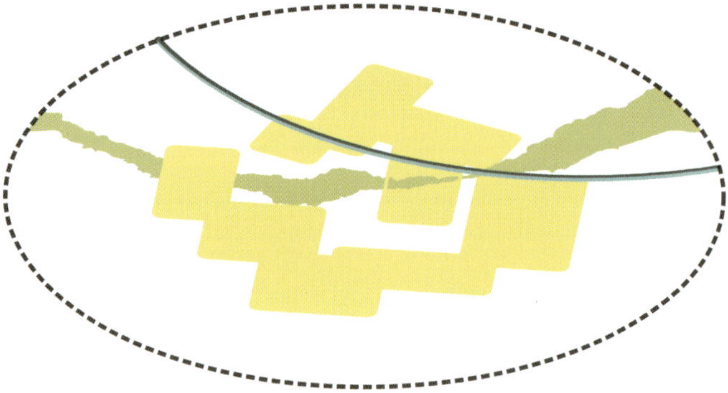

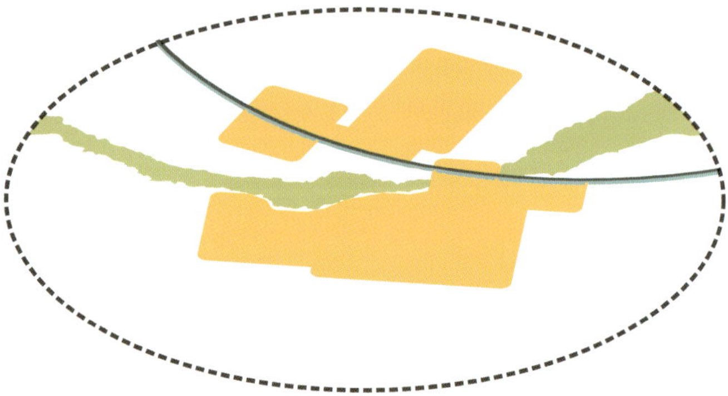

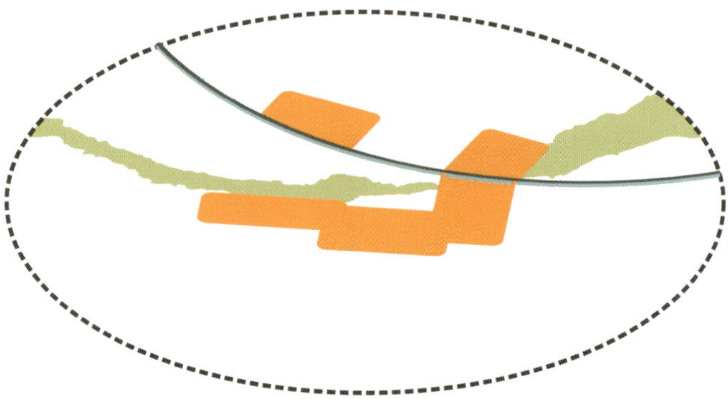

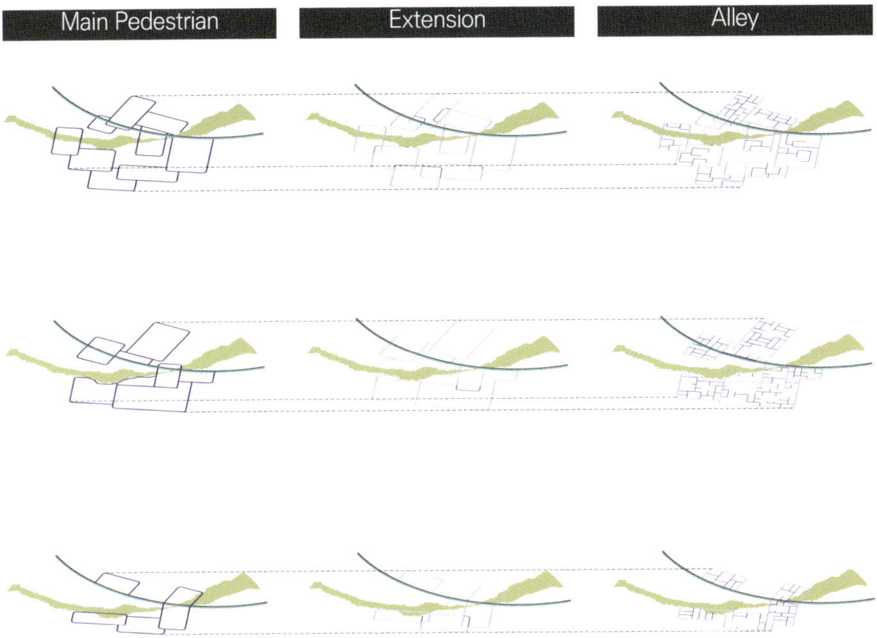

The eye level city's main purpose is to create walking-oriented means of transportation and high accessibility. So, we selected three hierarchies consisting of main pedestrian, extension, and alley so that people can move conveniently through walking on each mat.

1. Main Pedestrian
The Main Pedestrian road is located at the outer boundary of each mat. It serves as the main walking path for people and has the widest width of the three hierarchies.

2. Extension
The extension of each mat is located in a form in which each mat is overlapped.
Since the movement line using only the outer line can induce inefficient walking, a flexible movement line was created through the extension.

3. Alley
Alley is the way to enter each building from the mat.
It is the last step to induce people to walk quickly and conveniently by splitting the space inside the mat into small pieces.

The walking system of eye level city must follow the t-juction form.
Since wayfinding can be difficult due to narrow vision when walking, the t-junction form was established as the principle of walking to reduce collisions and make it easier for people to recognize their location.

1st Pedestrian way
2nd Pedestrian way
1st Pedestria
2nd

3rd Pedestrian way
3rd Mat
1st Pedestri

2nd Mat
3rd Mat
1st pedestrian way
3rd pedestrian way
2nd Pe

2nd Mat
3rd Mat
trian way

1st Pedestrian way
2nd Pedestrian way
1st Mat

3rd Mat
2nd Mat

3rd Pedestrian way
1st Pedestrian way

1st Pedestrian way

Bangkok Field Trip

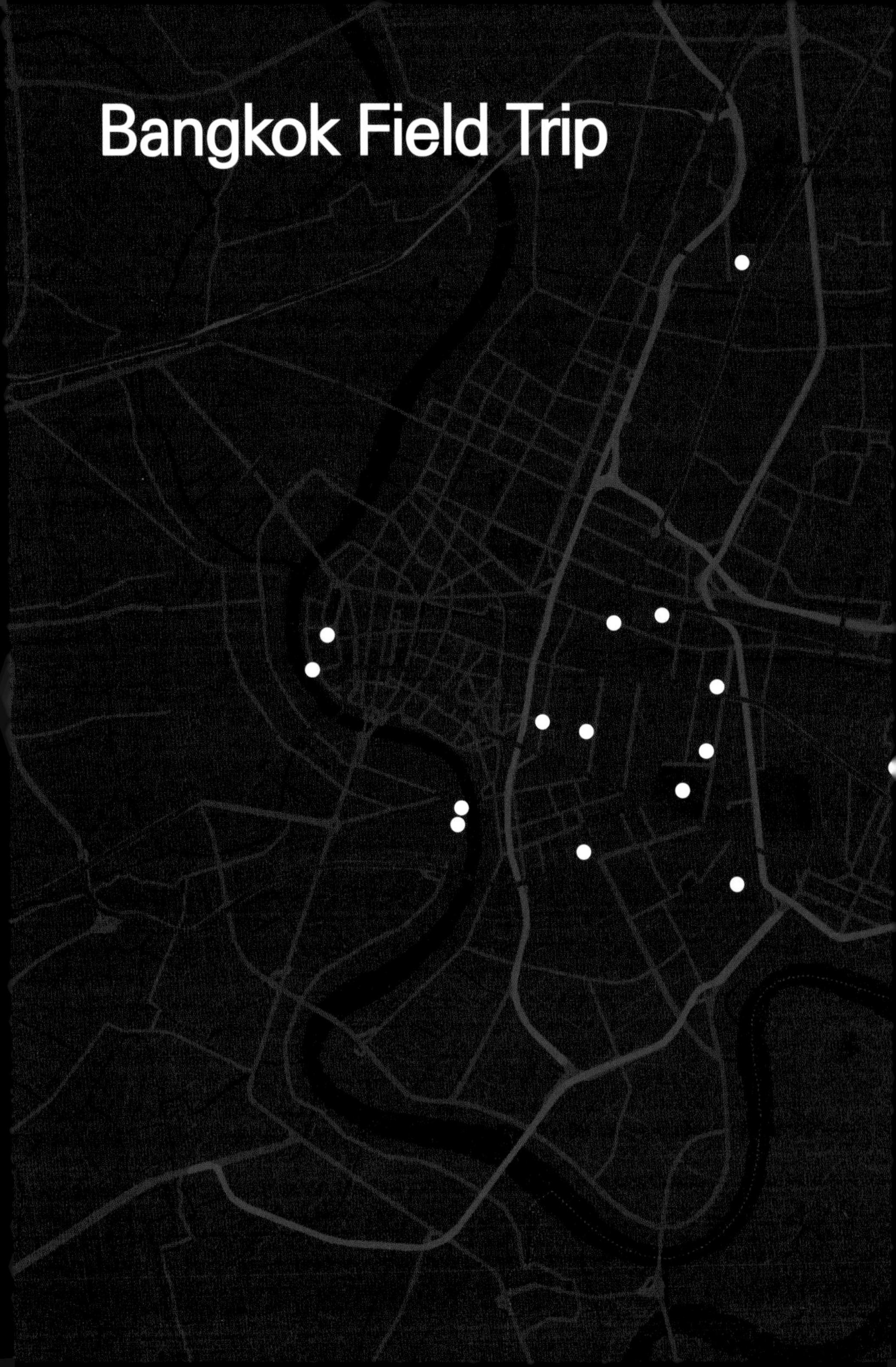

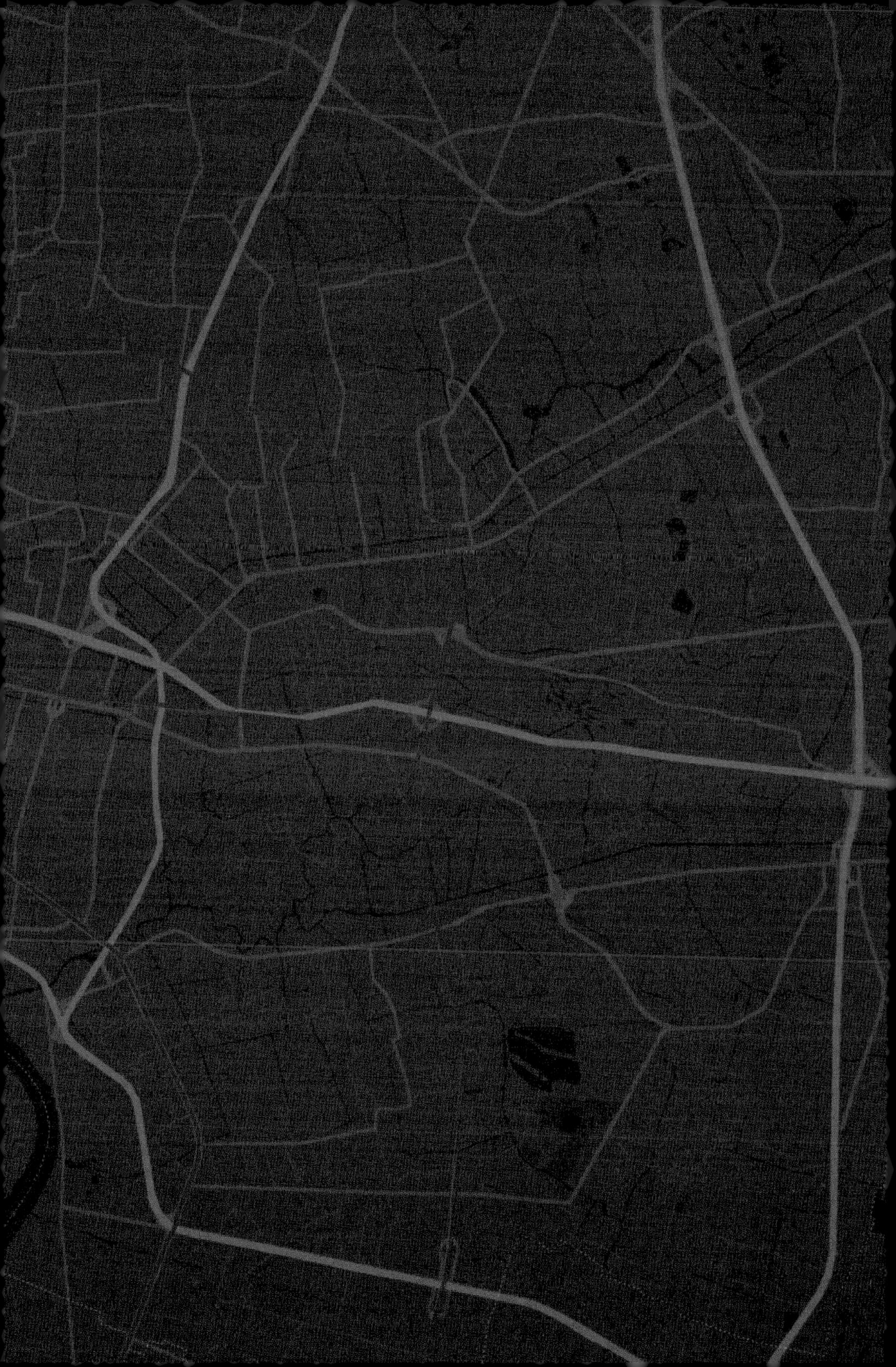

ICONSIAM
HERMÈS
HERMÈS

Exhibition

2024.07.02~2024.07.31 도만사

The Robust Compact City

미래강소도시

Seoul National University IPUD
2024-1 Urban Design Studio 1
with TAEJAE FOUNDATION

발행인
김기현
발행처
시공문화사 (Spacetime)
출판등록
1993년 3월 12일
주소
서울시 서대문구 독립문공원길 13 (03733) 극동프라자 5층
전화
02)3147-1212
팩스
02)3147-2626

http://www.spacetime.co.kr
spacetime@korea.com